Introduction to Discrete Event Simulation and Agent-based Modeling

T0192819

Theodore T. Allen

Introduction to Discrete Event Simulation and Agent-based Modeling

Voting Systems, Health Care, Military, and Manufacturing

 Springer

Dr. Theodore T. Allen, PhD
Integrated Systems Engineering
The Ohio State University
210 Baker Systems
1971 Neil Avenue
Columbus, OH 43210-1271
USA
e-mail: allen.515@osu.edu

ISBN 978-1-4471-5725-0   ISBN 978-0-85729-139-4 (eBook)

DOI 10.1007/978-0-85729-139-4

Springer London Dordrecht Heidelberg New York

British Library Cataloguing in Publication Data
A catalogue record for this book is available from the British Library

Cover design: eStudio Calamar S.L.

Printed on acid-free paper

Springer is part of Springer Science+Business Media (www.springer.com)

Foreword

The main purpose of this text is to provide an up-to-date foundation for applying discrete event simulation and agent-based modeling. It is perhaps true that no other book covers as many topics of interest for providing real-world decision-support including:

1. Open source simulation programming including Visual Basic (VB) and Net-Logo which provide inexpensive options for businesses,
2. Agent-based modeling,
3. Variance-reduction techniques such as Latin-Hypercube sampling,
4. Process improvement opportunity identification using the theory of constraints, lean production, and other contemporary methods,
5. Output analysis including selection and ranking and design of experiments,
6. Black box simulation and multi-fidelity optimization,
7. Quasi-Monte Carlo,
8. Subjectivity including the nature of probabilities and empirical distributions,
9. Input analysis including samples sizes and distributions fitting, and
10. Introduction to well-known software packages (ARENA and SIMIO).

As a result, this book is arguably more up-to-date than alternative texts for both research and practice. Also included are 100+ solved examples or problems.

This material has received good ratings when used in my introductory course in Integrated Systems Engineering at The Ohio State University. That course was (effectively) a single semester long and excluded the advanced content in Chaps. 8, 9, and 12 (variation reduction, VB, and agent-based modeling). It is believed that those chapters make the book relevant to both introductory undergraduate and graduate classes. While students at Ohio State have taken introductory probability theory prior to taking simulation, the intent is that prerequisites are not needed.

Acknowledgments

I would like to thank my wife Emily Patterson and my sons, Andrew and Henry, for support. Andrew helped to gather data for the supermarket project. I would like to thank my parents, George and Jodie, for additional support. In particular, Jodie edited the entire document. Austin Mount-Campbell provided content for the ARENA-related work and editorial assistance.

Muer Yang wrote much of the VB code and played the key role in establishing rigorous results related to election machine allocation. Douglas Samuelson contributed significantly to Chap. 12. Mikhail (Mike) Bernshteyn is my partner and respected colleague. Mike's coding and modeling have driven all of our elections related consulting work. David Sturrock helped to build and document a 3D-SIMIO model of the gourmet checkout aisles.

Gordon Clark and Allen Miller have both provided much needed support and encouragement. Also, I would like to thank Shane Henderson for ideas about election machine allocation and Mike Fry leading much of our research.

Fritz Scheuren and Steven Hertzberg provided the inspirational leadership that opened the world of election systems to Mike and me. Matthew Damschroder provided consistently thoughtful leadership for our Franklin County, Ohio projects with further leadership from the director, Michael Stinziano. Karen Cotton also provided key support. Ohio Secretary of State Jennifer Brunner and Antoinette Wilson continue to inspire us through their constant search for opportunities to take decisive, data-driven actions to aid Ohioans.

Tim Leopold and John Gillard supplied relevant information about simulation in practice including about embedded excel files. Michael Garrambone presented useful information about the uses of simulation and J. O. Miller clarified several issues about aggregation. The IIE Student Simulation Competition and Carley Jurishica and others from Rockwell International contributed. Edward Williams told me about modified Turing tests. Finally, the undergraduate students at The Ohio State University including especially Jonathan Carmona inspired the work and provided helpful feedback while Cathy Xia, Khalil-Bamoradia Kabiri, David Woods, and Suvrajeet Sen contributed many useful ideas.

Contents

Chapter 1
Introduction

Discrete event simulation and agent-based modeling are the subjects of this book. These types of simulation are merely two of many with others including systems dynamics, finite element analysis (FEM), and physical, human simulation. This latter type can involve running actual people through a scenario or game. Yet, discrete event simulation and agent-based modeling can offer natural approaches to help people think introspectively about their systems and realize efficiency gains. An animated version of the system being modeled is often a major outcome of the modeling activity. These animations can provide the best available way to engage untrained people in the application of all types of operations research and systems improvement activities.

This book begins with and focuses on the theory of discrete event simulation because it is the foundation for agent-based modeling also. It describes software in the last four chapters. Using standard software such as ARENA, AutoMod, NetLogo, ExpertFit, GPSS/H, ModSim, SIMIO, or WITNESS, it is possible to focus on software-specific details and forget the importance of theory. For a systematic review of related simulation software see Swain (2007). Yet, theory helps us understand how much data we need to feed into the software to obtain reliable results and how much and what type of results we need from the software to make defensible decisions. Without theory, the user risks being incompetent and generating untrustworthy, misleading results.

Also, as we will describe, real world problems can easily result in simulations that are too slow to provide all the insights that we need. For some of these situations, variance-reduction techniques and queuing theory are indispensable. For example, if we are trying to recommend numbers of machines needed for hundreds of voting locations, simulation of all of these subsystems and related optimization can be difficult or impossible. Queuing theory can provide transparent and defensible allocations with easy spreadsheet implementations.

Because discrete event simulation naturally estimates variation in its predictions, these methods are generally regarded as the starting point for modeling systems

T. T. Allen, *Introduction to Discrete Event Simulation and Agent-based Modeling,* DOI: 10.1007/978-0-85729-139-4_1, © Springer-Verlag London Limited 2011

involving high levels of uncertainty that cannot be ignored or "averaged over" in the results. We will have more to say about when finesse is possible and simulation is not technically needed. For example, there is generally uncertainty associated with systems of interest. Yet, linear programming that ignores these uncertainties can still be useful. Yet, even in many cases when randomness can be finessed, it can still be useful to build and display simulation models. The acts of making the model and inspecting the results have many intangible benefits. For example, the related activities can force teams to collect, analyze, and agree about data that are critical to the success of their organizations.

1.1 Domains and Uses

Paying customers for discrete event simulation and agent-based modeling are generally in several specific "application domains" or industrial sectors. A listing of these sectors together with typical questions includes:

- Manufacturing (How many machines and workers will we need?),
- Healthcare administration (How many machines and nurses will we need?),
- Call center support services (How many phone lines will we need?),
- Military applications (How many tanks will we need? How might incentives undermine an insurgency?), and
- Logistics (How many port access points will we need?).

Also, discrete event simulation and agent-based modeling have many uses. Examples are listed in Table 1.1 from capacity planning to test plan design.

The application of discrete event simulation to gaming and war gaming blurs the line between computer and human simulations because event simulations

Table 1.1 Possible applications of simulation and hypothetical questions to be addressed

Simulation applications	Example questions
Capacity planning	How many machines will be needed to meet a new order? How many airplanes will be needed to achieve strategic objectives?
Purchasing decision-making	How many regular and super-machines should we purchase? And how much raw material of different types will be consumed?
Project cost justification	In a six sigma project, if we reduce the defect rate to 2% will the reductions in inventory and rework cause the project to save money?
Strategic planning	What are the long term implications of a decision to use only air power in an attack plan?
Technology planning	If a machine has its processing or service time reduced by 20%, would the technology be worth the cost?
Training and gaming	How can I train my staff inexpensively? (There is really no clear line between discrete event simulations and gaming models for training)
Test plan design	What should the test specifications be to meet the performance requirements?

support war gaming and training. It has been reported that many areas of the US government have active discrete event simulation models and related activities. By one count, the US military has more than five major simulation models at each of the engagement, mission, and campaign levels.

The examples in the body of this book focus on voting systems. However, the problems at the end of each chapter and examples accompanying standard simulation software are designed to help students build the bridge from class concepts to their own future employment needs.

1.2 Questions about Voting Systems

Probably the main focus of discrete event simulation relates to predicting the properties of future queues or waiting lines. How long will they be if we use five machines? What if we use six machines? It is difficult to think of a higher stakes waiting line system than election systems. For example, in Allen and Bernshteyn (2006), we estimated that approximately 20,000 or more would-be voters did not vote because they were deterred by waiting lines in the Columbus, Ohio area alone. Worse perhaps was the fact that these voters were predominantly African Americans so that the election line systems were guilty of discrimination. Fortunately, by the 2008 election, we were able to help county officials put in place a transparent, simulation-theory-motivated voting machine allocation that resulted in minimal lines and no alleged discrimination (to our knowledge).

Related voting-systems questions constitute the set of motivating examples used throughout this book. It is hoped that, by using the same "newsworthy" example throughout, the reader will better understand how the concepts tie together. These questions are:

Question 1: How much time should one expect to spend (other than waiting) in voting?
Question 2: How long should voters expect to wait in total, roughly?
Question 3: What data inputs are needed to help officials?
Question 4: Which is better: full-faced machines or page-through machines?
Question 5: How many machines are required at each of 532 locations?

Each question corresponds to one of the chapters of this book. Specifically, the first question is a case in which simulation is not technically needed. Chapter 2 answers it highlighting the relationship between simulations and estimating expected values. Chapter 3 focuses on the requirements of an appropriate real-world data. Chapter 4 addresses the key application of discrete event simulation to forecasting expected waiting times and the issues associated with obtaining trustworthy, repeatable estimates. The answers relate to satisfying the conditions of the central limit theorem.

Next, Chapter 5 focuses on what we need from studying simulation model outputs to *prove* that one system option, within the range of validity of the simulation, is better than other system options. The focus here is on the problem of making multiple comparisons simultaneously, accounting for errors. Chapter 6 describes the queuing theory needed for answering some types of questions for which simulation is too cumbersome alone. For example, the fifth question involves decision-making at 532 locations.

Chapter 7 describes a real election systems simulation project and its possible relationship with class projects and exercises. It also briefly describes two management dogmas that may be helpful for generating alternative system designs to evaluate using simulation. These are the theory of constraints and lean production based on the Toyota Production System. Chapter 8, results are presented related to speeding up discrete event simulations, e.g., using so-called quasi-Monte Carlo and variance-reduction techniques including alternatives to ordinary pseudo random numbers.

The final three chapters describe software implementations of simulation. Chapter 9 focuses on custom code written using VB and requiring no licensing fees. Chapter 10 and Chapter 11 provide and introduction to a commercial package called ARENA. In Chap. 12, agent-based simulation and its history, the NetLogo programming language, and the relationship between alternative types of simulation and societal needs are described together with future directions.

1.3 Simulation Phases

Each organization tends to have its own nomenclature and way to conceptualize the phases of simulation. Here, the focus is on the terminology used by arguably the world's leading conference on discrete event simulation, i.e., the Winter Simulation Conference (WSC, www.wintersim.org). Therefore, we divide the endeavor of simulation modeling into five phases:

Phase 1: *Define* (who will do what? how roughly? and by when?),
Phase 2: *Input analysis* (collecting data and fitting the distributions needed as modeling inputs),
Phase 3: *Simulation or calculation* (creating and validating prediction models),
Phase 4: *Output analysis* (using validated models to compare alternatives and validate further),
Phase 5: *Decision support* (making charts, tables, and reports to help foster desirable choices).

Clearly, the process is often iterative with validation and output analysis requests coming back from management during Phase 5. Unlike the phases in a six sigma project, there is generally little discouragement for revisiting past phases other than the annoyance associated with communicating changing objectives inside a simulation team. Also, respected organizations such as Honda of America routinely perform a sixth phase in which they inspect completed projects to verify the extent to which historical predictions proved accurate.

Next, we focus on Phase 1 pointing out a few nontrivial aspects that have proved helpful to consider in past projects.

1.4 Phase 1: Define the System and Team Charter

In this phase, we determine what elements are within the scope of the simulation activity and out-of-scope. A flow chart or "workflow" is frequently used to show the sequence of operations. Workflows are often developed using software such as Microsoft® Visio®, Microsoft PowerPoint®, and/or using conventions systems such as those dictated by the international standards organization. In addition to the flow chart, typically a team charter is developed dictating who is responsible for the data collection of various types, model building, and report writing associated with the project. Such a charter is a contract between the team and management and can protect the team to some extent from "scope creep" in which the team is constantly tasked with more and more work.

As an example, consider the election systems problem of predicting the time that voters will need to register and vote in the next election. While this task might sound specific, it needs to be made clearer to become "actionable" by the team and for detailed data collection to begin. Project budgets are generally small and there is only so much that one can expect when planning to spend $10,000 or even $50,000. Hypothetically, imagine that we are offered a budget of only $10,000. Then, we might document that our goal is to predict the average (or, equivalently the mean) voting time of the voters in Columbus precinct 1A in the 2010 gubernatorial election. A related workflow is shown in Fig. 1.1.

In general, the define phase offers a chance for the technical team to interact with management such that resources for data collection can be budgeted taking into account both costs and timing. The team negotiates system outputs or "responses" (also called "measurable") and (potentially) targets for these outputs.

As a ball park estimate, consider that recruiting 60 representative voters from a county, bringing them in, and running a mock election on real direct recording equipment (DRE) voting machines requires approximately $12,000 in direct cost and an additional $8,000 in coordination, delivery, and supervision costs. This $20,000 does not include the cost for having analysts develop simulation

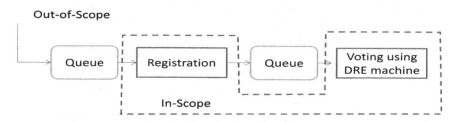

Fig. 1.1 A workflow for a hypothetical election systems simulation project

Table 1.2 A hypothetical charter for an election systems simulation project

Project	Predicting registration and voting times in a typical 2010 county precinct
Team members	John Doe, Li Wang, and Mary Smith
Timing	Final report planned for 6 weeks from project kick-off
Direct costs	$5,000 to recruit and time nine representative subjects
Responsibilities	John has the lead on reporting and Li has lead on the mock election
Primary objective	Predict the expected or mean voting time (the key response) of a randomly selected voter from the precinct
	Clarifying how this response varies as a function of the number of direct recording equipment (DRE) machines (the key factor) is of primary interest

models, paying simulation software license fees (e.g., ball-parked in at $50,000 per year for a full license of a powerful professional package), and having analysts develop final reports and present them.

In our real project with Franklin County, Ohio, the scope and budget were enlarged to include predictions at all locations and also responsibility for generating recommended numbers of machines at all locations. Also, we wrote our own simulation code using Visual Studio® and C++ to keep costs to a minimum and for computational speed, but our animation capabilities were also limited.

Table 1.2 shows the workflow for a miniature election systems project. As noted previously, the real project had a scope that included the entire county and additional responsibilities. Also, it is generally inadvisable to include sample sizes in planning below 20 data points. We discuss this "magic" number in Chap. 3.

1.5 Problems

1. What is input analysis?
2. What is (according to the chapter) the world's leading conference related to discrete event simulation?
3. What is output analysis?
4. What is a project charter?
5. A hospital is wondering about staffing for its third floor medical surgical unit floor which deals with medical problems suffered by former emergency department and post surgical patients. The trend is for the unit to treat 20% more patients of type 1 than in the previous year. Also, patients of type 2 arrive currently at the same rate as patients of type 2. The key inputs are the number of nurses (4 is current) and the hourly scheduling (2 work first shift, 2 s shift currently). Fifty percent of patients are discharged from med/surg. to home and the others are transferred to other areas of the hospital. Use this information to develop a work-flow (flow chart) and to clarify the goals and scope for a possible simulation project.
6. A call center has three main types of calls and 110 operators. Currently only 20% of the operators can handle the third type of call. They are considering

hiring additional skilled operators as well as training additional operators to be able to handle type three calls. Also, 2% of callers hang up while they are on hold and operating expenses exceed $10,000,000 per year. What are the factors and responses of potential interest for a simulation study?

7. You are given a $10,000 budget to gather information to try to make a certain cafe more profitable (one with long waiting lines). List two key responses and three possible controllable input variables (things you might change). Also, define the scope of your project in terms of which variables that you would time with a stop watch.

8. Write a charter for a hypothetical class project. Make sure to include objectives expressed in terms of at least two factors and two responses.

9. A manager of a successful call center service for banks is expanding operations and is trying to determine how many high-skilled and low-skilled associates to hire. Also, the center might switch to using more sophisticated call routing software at a license cost of $90 K per year. Highly skilled associates can handle all types of phone call questions while less-skilled associates can handle only about half of the types of calls. The center currently employs 75 associates with 38 being highly skilled. Its expansion comes in response to a new system at the banks that use their service, which is expected to increase the volume of mostly easy-to-handle calls by 30%. Assume workers cost $90,000 per year with benefits and management load. Explain in convincing detail how hiring you to build simulation models as a consultant would be profitable for the call center. If it is helpful, value your own time at $100 per hour.

Chapter 2
Probability Theory and Monte Carlo

In this chapter, the relationship between discrete event simulation and probability theory in general is described. Aside from creating insight-building animations, we will argue that discrete event simulation models are essentially calculators for estimating the "expected value" or "mean" of distributions. Therefore, reminding ourselves about the definition of the expected value is critical for comprehending simulation theory and the types of errors that arise in the practice of simulation.

This chapter focuses on the problem of predicting the registration and voting times in a future election. This problem is simple enough that it can be solved exactly without simulation. Therefore, it can be used to teach simulation for a case in which the true answer is known. Simulation estimates expected values with an error. Therefore, the example permits evaluation of the errors from simulation.

Further, the scope in the example here is the same as that from the previous chapter summarized by Fig. 1.1 and Table 1.2. By generating predictions for the expected times in three ways, the reader will also gain an appreciation for the "leap of faith" (LOF) and the associated concerns that are almost inevitably encountered in attempting to predict future events as well as the specific times at which the necessity for such leaps are typically encountered during the analysis process.

2.1 Random Variables and Expected Values

This section provides a review of elementary probability theory. (If the reader can confidently define expected values for continuous and discrete random variables, please skip to Sect. 2.2.)

We define a "random variable" (X) as a number whose value is not known at time of planning by the planner. While the value is unknown, generally the planner is comfortable with assuming a distribution function to summarize his or her

T. T. Allen, *Introduction to Discrete Event Simulation and Agent-based Modeling*,
DOI: 10.1007/978-0-85729-139-4_2, © Springer-Verlag London Limited 2011

beliefs about the random variable. If the random variable is discrete, i.e., it can assume only a countable number of values, then the distribution function is called a probability mass function, $\Pr\{X = x_i\}$ for $i = 1,...,n$. For example, X might represent the number of fingers (not including my thumb) that I am holding up behind my back. John Doe, a student, might have beliefs corresponding to:

$$\Pr\{X = 0\} = 0.1,$$
$$\Pr\{X = 1\} = 0.2,$$
$$\Pr\{X = 2\} = 0.3,$$
$$\Pr\{X = 3\} = 0.3, \quad \text{and} \tag{2.1}$$
$$\Pr\{X = 4\} = 0.1.$$

It is perhaps true that no one can tell John Doe that he is wrong in his current beliefs, although things that John might learn later might change his beliefs and his distribution. The above distribution has no name other than "discrete distribution" in that it is not Poisson or binomial (two famous discrete distributions). It is particular to John and his current state of beliefs. Yet, if John Doe declares the first 4 probabilities and then declares a value for $P\{X = 4\}$ other than 0.1, we might reasonably say that John Doe is incompetent in his ability to apply probability theory.

Next, for discrete random variables, the universally acknowledged definition for the mean or expected value is given by:

$$\text{mean} \equiv E[X] \equiv \sum_{i=1,...,n} x_i \Pr\{X = x_i\}. \tag{2.2}$$

For example, if we apply the distribution in Eq. 2.1 above, then the expected value is:

$$E[X] = (0)(0.1) + (1)(0.2) + (2)(0.3) + (3)(0.3) + (4)(0.1) = 2.1 \text{ fingers.} \tag{2.3}$$

Philosophically, the expected value is just as subjective as the distribution. It is not wrong to calculate an expected value and then to changes one's mind about his or her distribution. Yet, if one has a change of mind about the distribution in (2.1) logically (2.3) must be changed accordingly.

Similarly, if a random variable is continuous, it can (hypothetically) take on any value on at least some section of the real line. Continuous random variables are characterized by density functions also called "distribution functions" and written as $f(x)$. Distribution functions assign values proportional to the subjective likelihood of a random variable, X, achieving a value in the neighborhood of the value x. Just as appropriate or proper discrete density functions must have their probabilities sum to 1.0, appropriate continuous function distribution functions must have their values from $-\infty$ to ∞ integrate to 1.0.

As an example, consider that John Doe may state that his beliefs about the temperature in his home are characterized by the distribution function in Fig. 2.1.

Fig. 2.1 John Doe's density function describing his beliefs on room temperature

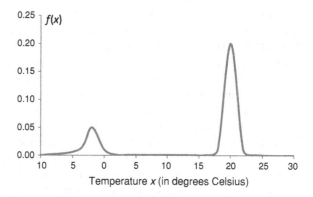

Temperature *x* (in degrees Celsius)

This distribution function does not resemble any frequently encountered, i.e., "famous" probability distribution and it is particular to John Doe at the time when he makes his declaration. It reflects his concern that the heater might break down, i.e., he believes that there is at least some chance of subzero temperatures.

The definition for the expected value of a random variable X given a continuous probability density function, $f(x)$, is defined as:

$$E[X] = \int_{-\infty}^{\infty} xf(x)dx \qquad (2.4)$$

A custom distribution as in Fig. 2.1 provides special challenges for estimating the expected value. This follows because there are only a finite number of cases for which we have the anti-derivatives to directly calculate the value in Eq. 2.4. There is no well-known anti-derivative for the function in Fig. 2.1. Custom distributions are important in discrete event simulation because generally the random variable whose expected value one is estimating does not resemble any frequently encountered continuous distribution. It might be the waiting time of a voter which is influenced by many factors such as machine breakdowns and the arrival rates of other voters.

A visual scan might yield an expected value of approximately 15°C, which appears to lie roughly at the center of mass of the distribution. In practice, we often attempt to find the closest "famous" distribution that roughly fits our beliefs. Using these famous distributions, we gain access to pre-established formulas relating the parameters that describe our assumed distributions to the mean and other properties of these distributions.

In this book, we focus on four well-studied or "famous" continuous distributions. These are the:

- **Uniform** (equally likely to be anywhere between a and b) written U[a,b],
- **Triangular** (must be greater than a, is most likely to be m, and must be less than b) written TRIA(a,m,b),
- **Exponential** (could be anywhere in the ball-park of $1/\lambda$) written EXPO($1/\lambda$), and
- **Normal** (somewhere within 3σ of the assumed mean μ) written N[μ,σ].

Note that much of the statistical literature writes the normal distribution as $N[\mu,\sigma^2]$, where σ^2 is the distribution variance. We choose to follow the excel "=NORMDIST()" conventions using the standard deviation instead of the mean. Also, the exponential is typically written in terms of its parameter λ and the reciprocal, $1/\lambda$, is the exponential mean or expected value.

We will have much more to say about each of these distribution functions. We will also describe so-called "empirical distribution" functions mainly for cases in which one has a large amount of data. In these cases, it might be assumed that the future data will be like the past data and not be limited by the shape of any famous distribution function. Obviously, predictions about important future events involve some degree of uncertainty. The distortion of approximating our true beliefs by one of the famous functions generally decreases the trust in our prediction process.

2.2 Confidence Intervals

Next, we review the standard approach to derive the confidence interval for the mean of a random variable based on data. Because these details are unusually important in the context of this book, even a reader with a good knowledge of elementary statistics, may want to read this section carefully.

In our notation, the data is written $X_1, X_2, \ldots, X_n$ where n is the number of data points. This approach creates an interval having a reasonably high and regulated probability of containing the true mean under specific assumptions given by the parameter α ("alpha"). The ability to create appropriate confidence intervals is considered essential to much of the simulation theory described in this book.

2.2.1 Confidence Interval Construction Method

Step 1. Calculate the sample mean (Xbar) using:

$$\text{Xbar} = (1/n) \sum_{i=1,\ldots,n} X_i \tag{2.5}$$

Step 2. Calculate the sample standard deviation (s) using:

$$s = \left\{ \left[\sum_{i=1,\ldots,n} (X_i - \text{Xbar})^2 \right] / (n-1) \right\}^{1/2}. \tag{2.6}$$

where $\{\}^{\frac{1}{2}}$ means take the square root of the quantity inside $\{\}$.

Step 3. Calculate the half width of the confidence interval using:

$$\text{Half width} = t_{\alpha/2,n-1} s / \left(n^{1/2} \right) \tag{2.7}$$

Table 2.1 Critical values of
the t distribution ($t_{\alpha,df}$)

df	α			
	0.01	0.025	0.05	0.1
1	31.82	12.71	6.31	3.08
2	6.96	4.30	2.92	1.89
3	4.54	3.18	2.35	1.64
4	3.75	2.78	2.13	1.53
5	3.36	2.57	2.02	1.48
6	3.14	2.45	1.94	1.44
7	3	2.36	1.89	1.41
8	2.9	2.31	1.86	1.4
9	2.82	2.26	1.83	1.38
10	2.76	2.23	1.81	1.37
20	2.53	2.09	1.72	1.33

and declare that the interval equals Xbar $\pm$ half width where typically $\alpha = 0.05$ and the value of $t_{\alpha/2,n-1}$ is found by consulting Table 2.1. In applying the formula in (2.7), with an overall value of $\alpha = 0.05$ and n data, we would look for the values with 0.025 and $df = n - 1$.

Step 4. (Optional). Check that that it is reasonable to assume that the individual data derive independent, identically distributed (IID) from a (single) normal distribution. If not, then do not trust the interval. Batching described in Chap. 4 might provide a useful way to derive trustworthy intervals for some cases.

Note that the symbol "df" stands for degrees of freedom, which has a geometric interpretation in the context of various statistical methods such as analysis of variance. It is merely an index to help us pick the right value from the table here.

Also, Step 4 is often not included in descriptions of confidence intervals but we will argue that this test is practically important in the context of output analysis in Chap. 5. Also, the details of the IID normally distributed conditions will be discussed at length in Chap. 4. In addition, Step 4 provides some motivation for the distribution fitting based methods described later in this section. This pertains to the voting systems example considered next.

Returning to our election systems example, imagine that the input analysis phase has progressed yielding the data in Table 2.2. These would hypothetically come from $n = 9$ registered voters from time trials with a stop watch in a mock election. Such a mock election would use ballots similar in length to what is our best projection for ballots in the 2010 gubernatorial election.

In Chap. 1, an example provided the team charter associated with this prediction problem (Table 1.2). In the context of this project, we are now in a position to complete phases 2, 3, and 4 simultaneously. We can derive a defensible prediction for the expected or mean sum of registration and voting times. With such a prediction, we can skip forward to Phase 5 (Decision Support).

If we use $X_1 = 7.4, ..., X_9 = 5.4$ and apply the confidence interval construction method, we will have our prediction. This is based on the assumption that the

Table 2.2 Registration and
direct recording equipment
(DRE) times in election
systems example

Person	Registration time(min)	Voting using DRE machine time (min)	Total (min)
Fred	0.2	7.2	7.4
Aysha	2.1	4.5	6.6
Juan	0.4	8.1	8.5
Mary	0.8	9.2	10
Henry	1.1	4.2	5.3
Larry	0.3	12.3	12.6
Bill	0.8	15.1	15.9
Jane	0.2	6.2	6.4
Catalina	0.6	4.8	5.4

future election will be similar to our mock election. Also, being good analysts, we will have error bars on our estimate. We will derive these from the confidence interval construction method applying the derived half widths.

Step 1. We calculate the sample mean (Xbar) using:

$$\text{Xbar} = (1/n) \sum_{i=1,\dots,n} X_i = (1/9)[7.4 + 6.6 + \cdots + 5.4]$$

$$= 8.68 \, \text{min.}$$

(2.8)

Step 2. We calculate the sample standard deviation (s) using:

$$s = \left\{ \left[\sum_{i=1,\dots,n} (X_i - \text{Xbar})^2 \right] / (n-1) \right\}^{1/2}$$

$$= \left\{ (7.4 - 8.68)^2 + \cdots + (5.4 - 8.68)^2 / 8 \right\}^{1/2}$$

$$= 3.58 \, \text{min.}$$

(2.9)

Step 3. We calculate the half width of the confidence interval using:

$$\text{Half width} = t_{\alpha/2, n-1} s / \left(n^{1/2} \right) = (2.31)(3.58) / \left(9^{1/2} \right)$$

$$= 2.75 \, \text{min.}$$

(2.10)

The interval is 8.68 ± 2.75 min or (5.92 min to 11.43 min).

Step 4. Figure 2.2 shows a histogram of the nine summed times. More precise methods to evaluate distributions quantitatively is described in Chap. 3. Such distribution testing is generally only trustworthy with 20 or more data points. Yet, here we merely say that the normal distribution is probably not a great fit. The observation of a second hump suggests that the true distribution might not be governed by a single bell shape. Strictly speaking, we know that the famous normal distribution is rarely (if ever) a perfect fit for a real process. However, in this case the fit is "extra" unreasonable.

Next, we describe two methods for generating predictions that do not depend on the assumption that the individual data are approximately normally distributed. Yet,

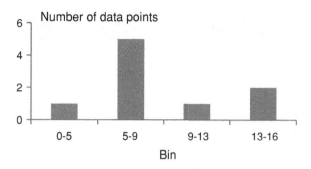

Fig. 2.2 Histogram for the registration plus voting time data showing two peaks

both of the methods that follow are associated with other strong assumptions that might give us concern. Despite the problem with normality, generating a confidence interval as shown in the above example is likely a defensible way to answer the problem stated. This explains why we say that simulation is probably not needed.

2.3 Expected Value Formula and Leaps of Faith

The method described in this section involves fitting famous distributions to the registration and voting times. The expected values are then calculated using the formulas associated with the famous distribution. Pencil and paper mathematics then permits the derivation of the forecast for the future expected registration time plus voting time.

The details of distribution fitting methods are the focus of Chap. 3. Here, let us assume that some software magically works through our data sets and fits triangular distributions to registration and voting times separately. Figure 2.3 shows the output from one such magical software package the Rockwell® Input Analyzer® which comes as a standalone in the same folder as the ARENA software. To develop this output using the Input Analyzer® one:

1. Opens the software and a new project using the File menu,
2. Puts the data in *.txt files perhaps using the NotePad built into Microsoft® operating systems, e.g., 0.2, 2.1,…,0.6 with each number in its own row and no commas or other separators,
3. Goes to File → Data File → Use Existing…, changes the "Files of type:" option to *.txt and selects the data file generated in the previous step, and
4. Selects "Triangular" from the "Fit" menu.

Ignoring the precise details temporarily, we now have a fit distribution. In our predictions of the future, let us entertain the assumption that registration times (in minutes) will come from a TRIA(0, 0.229, 2.29) distribution. Under this assumption, all times will be greater than 0 min (which makes sense), will have the most likely value of 0.229 min (we could live with that), and will be less than 2.29 min (an assumption that is somewhat limiting but may be acceptable).

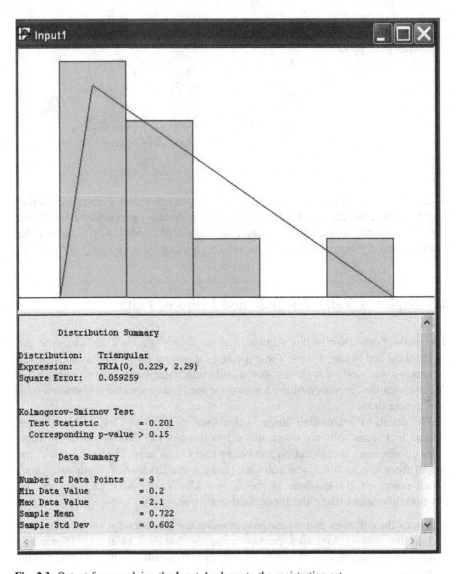

Fig. 2.3 Output from applying the Input Analyzer to the registration set

At this point let us take our leap of faith. From now on, we will play down the details of our nine real data points and simply assume TRIA(0, 0.229, 2.29) describes our beliefs. A similar process leads us to TRIA(4.0, 5.2, 16.0) min assumptions about future voting times, i.e., times voters need to use the direct recording equipment (DRE) machines once they are given access to their machines.

Having made our leap of faith, we are done with input analysis. We are ready for the calculation phase (Phase 3). In this phase, we are ready to gain one of the benefits of applying the famous triangular distribution. This benefit is that for parameters a, m, and b, the general formula for the mean of a triangularly distributed random variable is given by:

$$E[X] = (a + m + b)/3 \qquad (2.11)$$

We can also access the following general rule applicable to all pairs of random variables X_1 and X_2:

$$E[X_1 + X_2] = E[X_1] + E[X_2] \qquad (2.12)$$

This rule is general because it follows directly from the definitions of expected values in Eqs. 2.2 and 2.4.

In our example, we have X_1 is TRIA(0.0, 0.229, 2.29) and X_2 is TRIA(4, 5.2, 16). Again, when we made these assumptions, we can say that we made our "leap of faith" and "entered simulation land" where our input analysis data are irrelevant. Plugging the numbers into Eqs. 2.11 and 2.12, we see that our assumptions implied a predicted expected sum of registration and voting times is:

$$\begin{aligned} E[X_1 + X_2] &= (0.0 + 0.229 + 2.29)/3 + (4.0 + 5.2 + 16.0)/3 \\ &= 9.239666667 \pm 0.000000 \, \text{min} \end{aligned} \qquad (2.13)$$

Clearly, Eq. 2.13 is misleading. We know results are not infinitely trustworthy. However, we are in "assumption land" and our choice to entertain X_1 is TRIA(0.0, 0.229, 2.29) min and X_2 is TRIA(4.0, 5.2, 16) min has effectively caused these uncertainties to be ignored or irrelevant.

At least, our answer is not dependent on the data coming approximately from a single normal distribution. Also, next we will show how simulation in the same example adds a new type of "Monte Carlo simulation error" that makes the calculation in (2.13) look good in comparison. Generally, when one can apply calculus or probability theory to directly calculate an expected value, one should do it. Discrete event simulation merely estimates expected values with an error.

2.4   Discrete Event Simulation

This section introduces two key technologies. These are: (1) linear congruential generators (LCGs) and (2) inverse cumulative distribution functions. Together, these form the nontrivial components of by-hand discrete event simulations and permit simulation using spread sheets.

In the context of our election systems example, the application of these technologies turns out to derive a less desirable prediction for the expected times than those developed previously. However, the discrete event simulation methods introduced here have advantages in more complicated situations for which the previous methods (simple confidence intervals based on data and using calculus to derive expected values) are not applicable. So, for convenience, we introduction simulation technology in the context of an example for which we know the answer can be derived more accurately from probability theory or calculus.

A linear congruential generator (LCG) is a reasonably simple way to generate numbers that are "**pseudo random**" and associated with a uniform $a = 0$ and $b = 1$, i.e., U[0,1] distribution. We say that numbers are "pseudo random" because:

- They are not random in that we have a way to predict them accurately at time of planning and
- They closely resemble truly random numbers from the distribution in question.

There are pseudo random numbers of various levels of quality. Simulation trainers know that pseudo random numbers from LCGs are generally low in quality. This is because statistical tests can fairly easy show that they do not closely resemble actual U[0,1] random numbers. Yet, we use LCGs for instruction purposes because they illustrate the key concepts associated with pseudo random numbers.

The following method defines linear congruential generators (LCGs) in terms of three parameters: $a*$, $c*$, and $m*$. It is just a coincidence that we typically use a, c, and m when working with triangularly distributed numbers. This explains why we use the symbol "*" to clarify the difference.

2.4.1 Linear Congruential Generators

Step 1. Start with a seed, e.g., $Z_0 = 19$, and $i = 0$.
Step 2. $i \rightarrow i + 1$;

$$Z_i = \text{modulo}[(a*)(Z_{i-1}) + c*, m*],$$
$$U_i = Z_i/(m*), \quad \text{and}$$
(2.14)

where modulo is the standard function giving the remainder of $[(a*)(Z_{i-1}) + c*$ when divided by $m*$.
Step 3. Got enough? Yes, stop. No, go to Step 2.

In our example, we use $a* = 22$, $c* = 4$, and $m* = 63$. Yet, the above method defines an LCG for many combinations satisfying $a*$, $c*$, and $m* > 3$. The quality of the random numbers depends greatly on the specific choice with generally larger numbers spawning increasingly uniform seeming sequences of pseudo random U_i. The seed can also be important.

For example, Table 2.3 shows a sequence of 10 pseudo random uniformly distributed random numbers from an LCG. For completeness sake, the standard definition of a uniformly distributed random number from $a = 0$ and $b = 1$, i.e., U[0,1], is given by $f(x) = 1.0$ for $0.0 \leq x \leq 1.0$ and $f(x) = 0$ otherwise. Given this definition, does it seem reasonable to say that the numbers U_i for $i = 1,...,10$ in Table 2.3 resemble random numbers from a U[0,1] distribution?

Table 2.3	10 pseudo random	i	Z_i	U_i
U[0,1] numbers from a linear				
congruency generator		0	19	–
		1	44	0.698413
		2	27	0.428571
		3	31	0.492063
		4	56	0.888889
		5	39	0.619048
		6	43	0.682540
		7	5	0.079365
		8	51	0.809524
		9	55	0.873016
		10	17	0.269841

The answer depends on how accurate we are trying to be. Yes, they seem to superficially resemble uniformly distributed random numbers. However, continuing the sequence we observe cycling every 63 numbers. Consider that, in typical real problems, one uses 1 billion or more pseudo random U[0,1] numbers. Therefore, our LCG is not up to the task. We observe one type of departure from uniformity (cycling) after only 63 numbers. This occurs long before we get to the billion numbers that we need.

Industrial strength pseudo random U[0,1] number generators like "=RAND()" function in excel are far more complicated than LCGs. Yet, like LCGs they generate pseudo random numbers. Also, they have seeds, e.g., 19 in Table 2.3.

For the same seed, their sequence or "stream" is the same. The "=RAND()" function does not permit us to access the seed and it changes every time an Excel sheet field changes. However, if the "AnalysisToolPak" is added into excel, one can access the "Tools" → "Random Number Generator" feature. In more recent versions of excel, these options are available under Data → Data Analysis → Random Number Generation. Using whichever is appropriate to your version, one can generate streams of high quality pseudo random numbers of several types with adjustable seeds.

2.4.2 Inverse Cumulative Distribution Functions

Once we have pseudo random U[0,1] random numbers, it is generally of interest to convert them to pseudo random numbers of distributions of greater interest to us. For example, we might want pseudo random TRIA(0.0, 0.229, 2.29) numbers to generate plausible registration times for our simulated election. There are generally many approaches for converting a stream of pseudo random U[0,1] into numbers of the type that we desire. However, if we merely want a sequence of uncorrelated random numbers from a distribution of interest, a common and efficient approach is based on so-called inverse cumulative distribution functions, $F^{-1}(x)$.

Consider that the inverse cumulative distribution function for triangularly distributed random numbers is:

$$F^{-1}(u|a, m, b) = a + [u(m - a)(b - a)]^{1/2} \quad \text{for } u \le (m - a)/(b - a)$$

$$\text{or } b - [(1 - u)(b - m)(b - a)]^{1/2} \quad \text{otherwise} \tag{2.15}$$

where $[]^{1/2}$ means take the square root of the quantity in the brackets. Neglect temporarily how one derives this function. The key fact is that, once we have $F^{-1}(x)$, we simply plug in our pseudo random U[0,1] number as u and we derive a pseudo random number according to the distribution of interest.

For example, if we plug in $u = 0.698413$ and $a = 0.0$, $m = 0.229$, and $b = 2.29$ (in min), then Eq. 2.15 gives $F^{-1} = 1.07$ min. It can be checked that $u \le (m - a)/(b - a)$ so that we apply $b - [(1 - u)(b - m)(b - a)]^{1/2}$ in this example. This is our first pseudo random number according to the TRIA(0.0, 0.229, 2.29) distribution. It is not entirely trustworthy because it derives from an LCG, but it might seem plausible as a hypothetical registration time.

To better understand how inverse cumulative distribution functions work, consider the uniform distribution function, $f(x)$, its cumulative $F(x)$, and its cumulative inverse distribution function, $F^{-1}(x)$:

$$f(x) = 1.0 \quad \text{for} \quad a \le x \le b \quad \text{or 0.0} \quad \text{otherwise,} \tag{2.16}$$

$$F(x) = \int_{-\infty}^{\infty} f(z)dz = 0.0 \quad \text{for } x \le a, \tag{2.17}$$

$$(x - a)/(b - a) \quad \text{for} \quad a \le x \le b, \text{ and}$$
$$1.0 \quad \text{for } x \ge b, \text{ and} \tag{2.18}$$
$$F^{-1}(u) = a + (b - a)(u).$$

First, note that the inverse cumulative distribution function $F^{-1}(u)$ in Eq. 2.18 intuitively serves our purpose. If we plug in a number between 0.0 and 1.0, u, the result obtained lies between a and b. If u is closer to 0.0, then the result will be closer to a. If it is closer to 1.0, then the result will be closer to b. That is reasonable and desirable.

Also, consider how one can derive Eq. 2.18 from Eq. 2.17. Substitute $u = F(x)$ and solve for x. The result should give the right hand side in Eq. 2.18 for the relevant cases in which u is between 0.0 and 1.0. Finally, consider the plot of the cumulative distribution function, $F(x)$, as shown in Fig. 2.4. The starting numbers are equally likely to be between 0.0 and 1.0. Figure 2.4 shows a hypothetical value of 0.3000. Reading over and reading down gives a pseudo random number 0.3 fraction of the way from a to b.

The relationship between the inverse cumulative and generating pseudo random numbers is perhaps made clearer if we consider the custom distribution that John Doe hypothesizes or believes about the room temperature. Figure 2.5 shows how

Fig. 2.4 Plugging in pseudo random (PR) U[0,1] u's generates PR U[a, b] x's

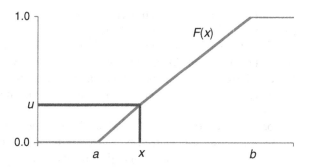

Fig. 2.5 Cumulative distribution function for custom temperature distribution

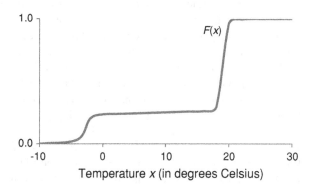

the slope of the cumulative is proportional to distribution function. John expects that the temperature will most likely be around 20°C but has a not insubstantial chance of being around −2°C. If one starts with a uniform pseudo random u on the vertical axis, one can see that it is almost certain that the result of reading over and down will either be around either −2° or 20°. Other values have very little chance (or "cross-section") to occur.

Other approaches for generating pseudo random numbers from specific distributions include the "acceptance-rejection method" which is useful even when F^{-1} is not available. In this method, variables are generated from one distribution, $g(x)$ from which it is easy to sample. Then, the numbers are conditionally eliminated based on a condition. Often, this condition relates to the ratio of probability density functions $f(x)/g(x)$.

2.4.3 Discrete Event Simulation

Now, we can generate pseudo random U[0,1] numbers using an LCG. We can also convert these numbers to pseudo random numbers from other distributions for which we have inverse cumulative distribution functions, $F^{-1}(u)$. We are ready to put the results together and generate discrete event simulations. Therefore, we will simulate all of the random numbers needed to complete one full event or replicate (sometimes call "replication") of interest.

The outputs from our simulations will be, in general, pseudo random numbers according to custom distributions whose mean values we are trying to estimate. We call these "discrete event" simulations because the quantities being simulated generally relate to occurrences happening at specific, identifiable times, i.e., events.

For example, consider a simulation of voters registering and voting. The scope of this simulation was defined previously in the define phase (Phase 1) to include only these two activities. In this fairly trivial scope, events associated with other voters and their interactions are irrelevant. A full replicate corresponds to the experience of a single simulated voter. The response of interest is the time elapsed between the event when a voter starts registration and the event when that voter completes voting using the voting machine. Here, one is trying to predict or estimate the expected value or mean of this elapsed time.

Table 2.4 shows five replications of discrete event simulation for the voter registration and voting prediction project. On the left-hand-side is the stream of pseudo random numbers from the LCG. Next, each number is transformed to a pseudo random time using the appropriate cumulative inverse distribution function, $F^{-1}(x)$. For this purpose Eq. 2.15 is applied. If the time is a registration time, then $a = 0.0$, $m = 0.229$, and $c = 2.29$ is used. If the time needed is a voting time, then $a = 4$, $m = 5.2$, and $c = 16$ is used. The resulting pseudo random number is not from any famous distribution. It is from the sum of two triangular distributions. This sum distribution has no special name.

Yet, it can be shown that the resulting stream of numbers, 8.491, ..., 7.788 are pseudo random independent identically distributed (IID) with an unknown true mean value equal to 9.239666667. This is the same number we derived previously using the exact formulas for the expected value. Since they are IID (more about this is discussed in Chap. 4) and from a distribution that is somewhat like a normal distribution, it is reasonable and appropriate to apply the confidence interval

Table 2.4 Discrete event simulation of 5 simulated voters registering and voting	I	Z_i	U_i	Simulated voter or replicate	Registration	Voting	Simulated time
	0	19	–		–	–	–
	1	44	0.698413	1	1.097	–	–
	2	27	0.428571	1	–	7.394	8.491
	3	31	0.492063	2	0.742	–	–
	4	56	0.888889	2		12.205	12.947
	5	39	0.619048	3	0.949		–
	6	43	0.682540	3		9.586	10.535
	7	5	0.079365	4	0.204		–
	8	51	0.809524	4		11.032	11.236
	9	55	0.873016	5	1.516		–
	10	17	0.269841	5	–	6.272	7.788

construction method from Sect. 2.2 to generate a range describing the mean. Remember, we are pretending that we do not know the mean equals 9.239666667 min and merely estimating it using our simulation results.

Discrete event simulation is one type of a more general form of statistical simulation called "Monte Carlo" simulation. The exact relationship is not important here. The name "Monte Carlo" derives from the city located near the south of France where gambling has historically occurred. Statistical simulation theory was often motivated by applications relating to gambling including gambling in Monte Carlo.

Our general formula for Monte Carlo estimates for expected values is:

$$\text{Monte Carlo estimated expected value} \equiv \text{Xbar} \equiv \text{the sample average} \quad (2.19)$$

with Monte Carlo errors estimated using the half width from our confidence interval. In our example, the sample average equal Xbar equals 10.2 min with sample standard deviation (s) equal to 2.1 min. The half width is 1.9 min from Eq. 2.7. Therefore, the Monte Carlo simulation estimate can be quoted as 10.2 min ± 1.9 min.

Note that the error for Monte Carlo estimates in Eq. 2.19 declines according to the number of simulation replicates, n, that we choose to do. The exact proportionality is given by our confidence interval half width Eq. 2.7 as $1/n^{1/2}$ or the reciprocal of the square root. This means that the simulation error is directly attributable to our failure to simulate additional replicates. If we run more replicates, we reduce the error. Yet, in some cases of interest, replication times can consume more than 1 h. In those instances, we will generally need to live with large half widths. This is not true, however, for our spread sheet simulations such as the one in Table 2.4. In this case, it takes virtually no time to copy formulas down and perform 10,000 replicates. Yet, in this case, we are limited by the poor quality of our LCG.

2.5 Monte Carlo Errors

The difference between our Monte Carlo estimate, Xbar, and the true mean, which we might denote E[X], is an error. We define the Monte Carlo error as simply:

$$\text{Monte Carlo error} \equiv E[X] - \text{Xbar}. \quad (2.20)$$

From our discussion of confidence intervals, we know that we can drive this error to near zero by performing sufficient replicates. Consider that, in general, simulation is performed using pre-made software with easily adjustable numbers of replicates. Therefore, the only excuse for not driving the Monte Carlo errors to near zero is a function of the time each replicate requires from the computer processors and the patience of the human analysts who are awaiting the results.

The Monte Carlo error is in addition to any error we might imagine stemming from an imperfect input analysis phase. In our example, we know we only had nine data points. As a result, our results cannot be particularly trustworthy. Our leap of faith is making us concerned and the Monte Carlo error only adds to that concern.

In conclusion, the particular simulation example described here happens to permit us to directly estimate the Monte Carlo error. The true mean under our assumptions, E[X], we know from direct calculation is 9.239666667 min. The Monte Carlo estimate is Xbar = 10.2 min. Therefore, the Monte Carlo error from Eq. 2.20 equals −1.0 min. Generally, we will not know our Monte Carlo errors and need to estimate or bound them using half widths from our confidence intervals.

2.6 Monte Carlo Simulation Example

2.6.1 Problem

Consider a pseudo random number X that is assumed to be TRIA(1, 5, 12). Also, consider another random variable, Y, that is either X or 4, whichever is greater. We can write: $Y = Maximum(X,4)$. Use Monte Carlo simulation to estimate $E[Y^2]$. Perform sufficient numbers of replicates until the half width of your Monte Carlo estimate is less than or equal to 11.

2.6.2 Solution

The overall strategy is to generate pseudo random numbers Y^2 and to use the sample mean to estimate the true mean, i.e., apply Eq. 2.19. Since we are not explicitly asked to apply an LCG, we will not use them. Instead, we will apply the higher quality pseudo random numbers from Excel-based on Tools → Random Number Generation → Uniform random numbers or, in some more recent versions, Data → Data Analysis → Random Number Generation → Uniform. We set the seed equal to 1. We use the formulas shown in Fig. 2.6.

The spreadsheet shows how the process starts with pseudo random numbers in cells A5 and below. Then, the inverse cumulative is used to generate triangularly distributed pseudo random numbers, Y values, and Y^2 values successively. The distribution of Y^2 has no famous name but we can estimate its mean using the sample average. After 10 replicates, the Monte Carlo estimate is 44.4 but the half width is 16.97, which is too large. Therefore, we apply 20 replicates. This yields a Monte Carlo estimate for the mean equal to 36.7 and a half width of 10.7. This is good enough for our purposes. Note that the "=TINV" function has been applied. This function requires that we multiply our alpha values by 2.0 to obtain the standard critical values that we need.

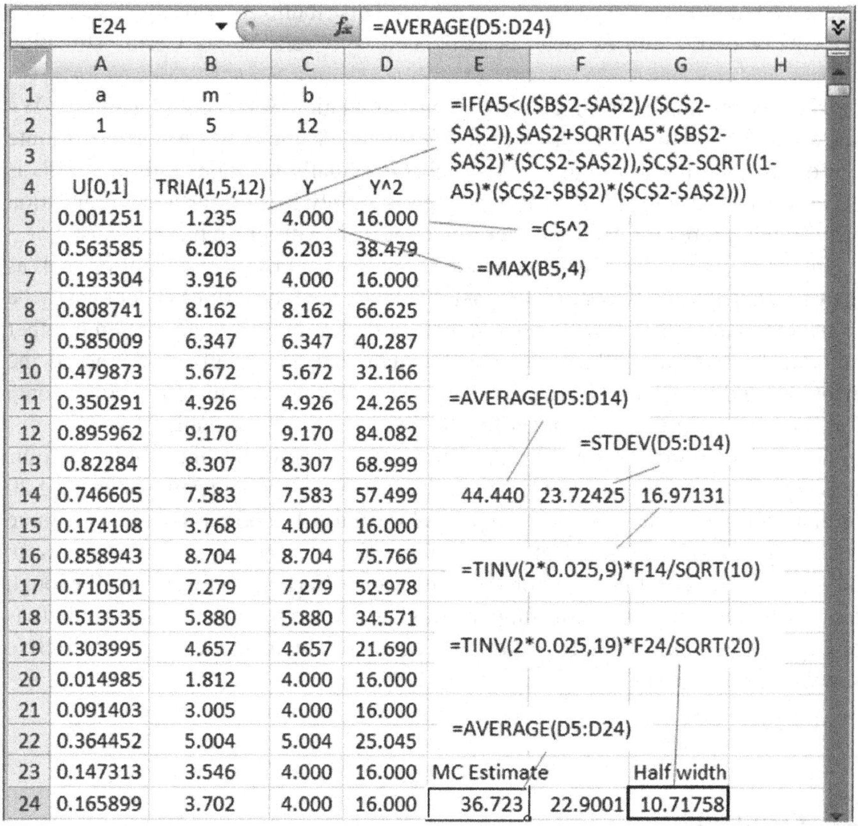

	A	B	C	D	E	F	G	H
	E24			f_x =AVERAGE(D5:D24)				
1	a	m	b		=IF(A5<(($B$2-$A$2)/($C$2-			
2	1	5	12		$A$2)),$A$2+SQRT(A5*($B$2-			
3					$A$2)*($C$2-$A$2)),$C$2-SQRT((1-			
4	U[0,1]	TRIA(1,5,12)	Y	Y^2	A5)*($C$2-$B$2)*($C$2-$A$2)))			
5	0.001251	1.235	4.000	16.000	=C5^2			
6	0.563585	6.203	6.203	38.479				
7	0.193304	3.916	4.000	16.000	=MAX(B5,4)			
8	0.808741	8.162	8.162	66.625				
9	0.585009	6.347	6.347	40.287				
10	0.479873	5.672	5.672	32.166				
11	0.350291	4.926	4.926	24.265	=AVERAGE(D5:D14)			
12	0.895962	9.170	9.170	84.082		=STDEV(D5:D14)		
13	0.82284	8.307	8.307	68.999				
14	0.746605	7.583	7.583	57.499	44.440	23.72425	16.97131	
15	0.174108	3.768	4.000	16.000				
16	0.858943	8.704	8.704	75.766	=TINV(2*0.025,9)*F14/SQRT(10)			
17	0.710501	7.279	7.279	52.978				
18	0.513535	5.880	5.880	34.571				
19	0.303995	4.657	4.657	21.690	=TINV(2*0.025,19)*F24/SQRT(20)			
20	0.014985	1.812	4.000	16.000				
21	0.091403	3.005	4.000	16.000				
22	0.364452	5.004	5.004	25.045	=AVERAGE(D5:D24)			
23	0.147313	3.546	4.000	16.000	MC Estimate		Half width	
24	0.165899	3.702	4.000	16.000	36.723	22.9001	10.71758	

Fig. 2.6 Microsoft® excel used to estimate the mean of a random variable

2.7 Voting Systems Example Summary

Collecting results from various methods, we first forecast the expected voting time to be 8.7 ± 3.6 min. This estimate involves a minimal leap of faith because we simply applied a confidence interval from the original input analysis data. Yes, the original data were not normally distributed to a good approximation (two humps). Still, the interval does reflect with some appropriateness the limitations of our nine subject data set. As a result, it is probably the best answer to the original forecasting question described here.

Our next interval derived from fitting distributions to the data. Then, standard formulas from probability theory were applied to estimate a mean of 9.239666667 ± 0.000000. We know that the precision in this estimate is misleading. The ±0.000000 does not reflect the leap of faith that we made after concluding our input analysis when we picked the two TRIA distributions.

Then, we put blinders on and took our assumed distributions seriously. Yet, at least the 9.2239666667 min estimate is associated with zero Monte Carlo estimation error.

The last estimate, from discrete event simulation was 10.2 ± 1.9 min. We know that this estimate is the worst. The predicted mean has a nonzero Monte Carlo error and the uncertainty (± 1.9 min) is actually an under estimate. This is because the ± 1.9 min estimate error bound is simply Monte Carlo or "replication error" and ignores the additional errors associated with our input analysis.

In our real simulation project for the county, we focused on quantities such as the expected waiting times of the worst precincts. Such quantities cannot easily be estimated in any other way besides Monte Carlo discrete event simulation or other numerical techniques. Therefore, we did make a leap of faith and ignored the errors related to our input analysis. Also, we could not calculate our true Monte Carlo errors since exact formulas for the mean were not available.

We simply estimated bounds on our errors using confidence interval half widths (based on Eq. 2.7). We tried to keep the errors to a reasonable level by applying 20 or more full replications. Usually, in research we use 10,000 replicates to get three decimal points of accuracy. But in the real election systems case, simulations were far too slow to permit that. Fortunately, the Monte Carlo errors were small enough for providing helpful decision support. In Chap. 4, we focus on other simulations of expected waiting times for additional cases in which Monte Carlo or discrete simulations are needed.

2.8 Problems

1. What is a random variable?
2. What is an expected value of a random variable?
3. What is a linear congruential generator (LCG)?
4. Why do we generally avoid using LCGs in addressing real world problems?
5. What is the "sample standard deviation"?
6. What is a "half width"?
7. What are Monte Carlo errors?
8. Consider the following data: 2.5, 9.2, 10.2, 9.8, 9.2, 10.3, 10.2, 2.8, and 10.1. Develop a 95% confidence interval for the mean assuming that the data are approximately normally distributed.
9. Consider the measured registration times 1.0, 2.4, 2.0, 3.5, and 1.4. Develop a 95% confidence interval for the mean using the given t-table.
10. Comment on how reasonable it is to assume that the data in problem 8 derive from a single, normal distribution.
11. Consider the measured voting times 5.0, 7.0, 9.0, 5.4, 3.0, and 4.4 with sample mean 5.6 and sample standard deviation 2.1. Develop a 95% confidence interval for the mean using the given t-table.

12. Considered the outputs from different replicates of a simulation given by 22.1, 18.3, 25.7, and 22.8 (waiting times in minutes). Give the Monte Carlo estimate for the mean waiting time and its half width.

13. Assume X is distributed according to $f(x)$, and 10.1, 19.4, and 23.0 are pseudo-random numbers from $f(x)$. Also, assume $\mu = \int_{-\infty}^{\infty} xf(x) = 19.0$. Estimate as accurately as possible $E[X + 3X]$ and $\mathrm{Var}[X]$. Estimate the errors of your estimates.

14. Assume X is distributed TRIA(4, 9, 10). Estimate $E[X^2]$ using Monte Carlo simulation.

15. Assume X is distributed according to $f(x)$, and 9.1, 20.3, 19.4, and 23.0 are pseudo-random numbers $f(x)$. Also, assume $\int_{-\infty}^{\infty} xf(x) = 22.0$. Estimate as accurately as possible $E[2X]$ and $E[X^2]$. Estimate the errors of your estimates.

16. Assume that X is U[10,25], what is $E[X]$? Estimate the answer using probability theory and also Monte Carlo simulation.

17. Assume X is triangularly distributed with parameters $a = 2$ h, $b = 10$ h, and $m = 3$ h. What does this assumption imply about $E[X]$? Also, describe this assumption in one or two sentences using everyday language.

18. Assume that X is U[10,25], what is $E[X^2]$? Estimate the answer using Monte Carlo simulation. Make sure your half width is less than 1.0.

19. Assume X is triangularly distributed with parameters $a = 2$ h, $b = 10$ h, and $m = 3$ h. Generate three pseudo random triangularly distributed random variables using the inverse cumulative and the uniform pseudorandom numbers 0.8, 0.3, and 0.5.

Chapter 3
Input Analysis

This chapter describes methods for gathering and analyzing real world data to support discrete event simulation modeling. In this phase, the distribution approximations for each process including arrivals are estimated using a combination of field observations (e.g., based on stopwatch timings) and assumption-making. In many cases, the time and cost of input analysis will actually exceed expenses from all other phases. It may also be necessary to put instrumentation into place to provide the accurate time measures, greatly delaying the entire project.

Section 3.1 describes simple strategies to help ensure that a sufficient amount of data is collected. Next, Sect. 3.2 describes approaches based on relative frequency histograms and the sum of squares error (SSE). These procedures, like the Rockwell® Input Analyzer, develop recommended choices for distributions but with some arbitrariness. Section 3.3 describes the rigorous Kolmogorov–Smirnov (KS) hypothesis testing method for ruling out distributions. The KS method avoids arbitrariness by using so-called empirical cumulative distribution functions which are also defined. Section 3.4 contains a numerical example that reviews related methods and concepts.

3.1 Guidelines for Gathering Data

In general, input analysis comes after the project is defined and the charter is made. Yet, planning for the data collection in the input analysis phase is sometimes done in the define phase, e.g., as part of the charter. This might occur because the expenses for data collection and the related timing might constitute a major part of the project cost. A natural starting point in planning these costs and the input analysis phase is a flowchart showing the subsystem of interest. This could be based on imagination or on thorough observation of the system being studied or a

T. T. Allen, *Introduction to Discrete Event Simulation and Agent-based Modeling*,
DOI: 10.1007/978-0-85729-139-4_3, © Springer-Verlag London Limited 2011

combination of both. Evaluation of the costs of collecting data for every process in the flowchart could, hypothetically, cause the team to limit (or expand) the project scope and goals.

The number 20 is a reasonable minimum number of measurements that can allow distribution fitting and defendable model development. We would generally want to measure at least 20 service times and, in some cases, 20 interarrival times for every bank of servers in our model. For example, if we were modeling a supermarket checkout, we might assume that the interarrival times are exponentially distributed and only measure 20 service times on a single of three checkout aisles. We might assume that the other two aisles have the same service distribution.

As justification, consider that the number 20 is the standard number for initial evaluation in so-called indifference zone (IZ) procedures which seek to aid in ranking system alternatives in Chap. 5. Also, in standard statistics process control (SPC) for monitoring generic processes, 25 subgroups are the world standard number for initial system characterization (Allen 2010). Frequently, some of these subgroups are removed from consideration and estimation is based on approximately 20 subgroups.

Timing interarrivals can be important if we are not content in assuming a constant exponential distribution. For example, in our voting project we might time n equals 100 interarrivals so that we can check the shape of the distribution including whether the arrival rate differed over the day predictably.

Another consideration is the possibility of *rare events* with potentially critical impacts on the responses of interest. Assume our initial estimate is that the rare event might characterize p_0 fraction of the entities under consideration. For example, from historical experience we guess that approximately $p_0 = 0.01$ fraction of voters have severe handicaps that affect their voting times. From studying the negative binomial observation, we can estimate the number of entities, n, we need to sample to observe r rare events (assuming independent trials). The standard formula for expected number of trials or sample size is:

$$E[n] = (r)(1 - p_0)/(p_0).   \qquad (3.1)$$

Ideally one would budget for sufficient trials such that $r = 20$ rare events can be observed. Yet, this might often be prohibitively expensive in both time and money. With $r = 5$, the discrete count of rare events might be roughly approximated by a normal distribution permitting the development of standard confidence intervals on the actual value of p_0 to be derived from the data. In low consequence situations, e.g., student projects, planning for $r = 1$ entities to be timed might be reasonable.

As an example, consider the problem of timing voters with the plan to include $r = 20$ handicapped persons. Then, we would need to time $(20)(1 - 0.01)/0.01 = 1,980$ voters to observe an expected 20 handicapped ones. This is approximately the number of voters we did time in our real project for Franklin County. Fortunately, these service times were recorded and documented in previous elections. The officials took the time to retrieve these earlier data, remove any voter identification, and send them to us in a spreadsheet.

A reasonable alternative approach would have used access to mock systems for experimentation. In such a scenario we would directly identify 20 entities or individuals associated with the rare event. Then, we could directly time only these 20 on our mock equipment. This would have lower fidelity than observing the actual system but it could save the cost of timing more than a thousand voters by hand. We might also have been able to identify the fraction, p_0, accurately from historical data, e.g., the number of logged requests by voters for special assistance.

Figure 3.1 shows a hypothetical sampling plan to support an election systems simulation project. The plan to collect 100 interarrival times reflects our subjective concern about applying a Poisson process, i.e., a single fixed exponential distribution. In manufacturing or with scheduled arrivals, Poisson processes are seldom relevant. For example, parts tend to arrive at machines at approximately fixed intervals and a normal distribution with a standard deviation much smaller than the mean might make sense. In other cases arrivals are uncoordinated; however, it is probably not as important to check that the interarrival times are approximately exponentially distributed. Uncoordinated arrivals are generally well approximated by exponential distributions.

It can be important, however, to identify whether arrivals come in groups or batches ("batch arrivals"). Also, it can be important to check whether the average interarrival time is changing over time in a predictable way, i.e., the gaps are "nonhomogeneous" or uneven. For example, there might be more customers arriving in predictable lunch "rush-time" than in other periods. Chapter 4 and Sect. 10.4 discuss modeling with these complications.

The plan to collect only 20 registration times reflects our confidence that the registration process is generally stable and not a critical "bottleneck" that determines system performance. The plan to collect 1,980 voting times reflects our desire to observe 20 handicapped voters and our confidence that these times can be derived from existing instrumentation with reasonable cost. Note that, in a real election systems context, the registration process might also be affected by other important rare events, e.g., individuals with limited documentation.

Consider also that the sampling plan might change depending on the allowed time for the project. For example, a project designed to reduce the loss of customers during rush periods might limit data collection to the length of the rush period. For example, lines might form only between 11 am and 1 pm. A different project designed to cut staffing costs during non-rush periods might require more sampling because of the greater length of the non-rush periods.

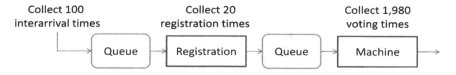

Fig. 3.1 Sampling plan for a single precinct simulation project

3.2 Relative Frequency Histograms and SSE

Once the data have been collected, a natural first step is to create histograms. Histograms provide bar chart visualizations of the shapes of the distributions. Also, least squares estimation described in this section based on histograms is often the primary method for estimating distribution parameters. The derived SSE from these fits is often used to motivate the choice of one distribution over another for a given process.

3.2.1 Relative Frequency Histograms

The following method creates histograms with equal length "bins" or intervals. The histograms summarize visually the dataset, X_1, X_2, ..., X_n.

Step 1. Calculate the number of "k" equal length "bins" or intervals. If the number of data points is n, we might use Sturges' rule to estimate the number of bins:

$$k = \text{roundup}[\log_2(n) + 1] \qquad (3.2)$$

where "roundup" simply means round the fraction up to the nearest integer and $\log_2()$ refers to the logarithm function to the base 2.0. Results from Eq. 3.2 are shown in Table 3.1.

Step 2. Calculate the endpoints of the bins, $q(i)$, using:

$$q(i) = \text{minimum}(X_1, X_2, \ldots, X_n) + (i)[\text{maximum}(X_1, X_2, \ldots, X_n) \\ - \text{minimum}(X_1, X_2, \ldots, X_n)]/(k) \qquad (3.3)$$

for $i = 0, \ldots, k$. In some cases, it is reasonable to adjust the $q(i)$ manually to hit round numbers to make the plot look easier to interpret. This arbitrariness in the bin rules can affect model fits and comparisons, however.

Step 3. Count or tally the number of observations or frequency in each bin, $C(i)$ for $i = 1, 2, \ldots, k$ and the relative frequencies, $rf(i)$, using:

$$rf(i) = C(i)/(n) \text{ for } i = 1, 2, \ldots, k. \qquad (3.4)$$

Table 3.1 Number of bins as recommended by Sturges' rule for histograms	n	# bins = roundup[1.442 ln(n) + 1]
	10	5
	20	6
	50	7
	100	8
	200	9
	1,000	11
	10,000	15

Fig. 3.2 Relative frequency histogram for election systems example and associated continuous distribution where the vertical axis can be interpreted as the relative frequency and probability

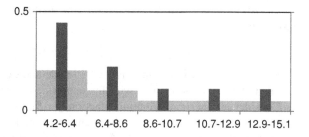

For points on the bin boundaries (if any), count them in the higher bin. Count the lowest point and highest points in the first and last bins respectively.

Step 4. Bar chart the relative frequencies versus the bin definitions. Here, we use the $q(i - 1) - q(i)$ notation to describe bin i for $i = 1,...,k$.

As an example, consider the hypothetical $n = 9$ voting times: 7.2, 4.5, 8.1, 9.2, 4.2, 12.3, 15.1, 6.2, and 4.8. Therefore, $X_1 = 7.2$, $X_2 = 4.5$,... Sturges' rule from Table 3.1 gives $k = 5$ bins. Then, the bin endpoints are $q(0) = 4.2$, $q(1) = 6.4$, $q(2) = 8.6$, $q(3) = 10.7$, $q(4) = 12.9$, and $q(5) = 15.1$. The relative frequencies are then: $rf(1) = 0.4$, $rf(2) = 0.2$, $rf(3) = 0.1$, $rf(4) = 0.1$, and $rf(5) = 0.1$. The resulting bar chart is shown in Fig. 3.2.

Note that the above approach is subtly different from the method in Microsoft® Excel. The Excel histogram method under Data Analysis → Histogram uses fewer bins for small data sets. Also, if custom bin widths are applied, the counts are based on values below or equal to the lower bin bounds. Excel also labels the bins using the lower bounds which might seem misleading.

Note also the grey boxes in Table 3.1. Every relative frequency histogram can be used to generate an associated continuous distribution. All values within each bin are assigned a value equal to the relative frequency multiplied by a scale factor given by:

$$\text{Scale} = \frac{\#\text{bins}}{[\text{Maximum}(X_1,...,X_n) - \text{Minimum}(X_1,...,X_n)]} \quad (3.5)$$

so that the scale factor makes the area under the associated continuous distribution equal to 1.0, i.e., the distribution $f(x) = (\text{Scale}) \times [rf(i)]$ for x in bin i is "proper."

3.2.2 Sum of Squares Error

After the relative frequency histogram has been constructed, the distributions are fitted using the least squares error. In some implementations, these fits are constrained so that the fitted distributions have pleasing properties. In this section, methods are described to fit distributions to relative frequency histograms that minimize the SEE.

To estimate the best fit model, we first define the midpoints, $m(i)$, of the histogram bins using:

$$m_i = [q(i - 1) + q(i)]/(2) \quad \text{for } i = 1, \ldots, k \tag{3.6}$$

where $q(i)$ is the ith bin limit. Also, we include additional cells below the lowest cell and above the largest cell with frequencies equal to 0.0.

Consider a generic distribution function, $f(x, \boldsymbol{\beta})$ with a vector of parameters $\boldsymbol{\beta}$ that we are trying to estimate. For example, triangular distributions have three parameters not including the area $\boldsymbol{\beta} = (a, m, b)'$. All parameters are estimated by solving a sum of squared error estimation problem:

$$\boldsymbol{\beta}_{est} = \text{argmin}\left\{ SSE = \sum_{i=1,\ldots,k} [(\text{Scale})rf(i) - f(m_i, \boldsymbol{\beta}_{est})]^2 \right\}. \tag{3.7}$$

where argmin{ } refers to the argument that minimizes the SSE. In words, the parameter values we estimate are the ones that make the fit distribution function pass through the histogram points. Often, it is reasonable to apply constraints in the optimization in Eq. 3.7.

For an example, we return to the election systems data set and histogram in Fig. 3.2. Also, we will fit the triangular distribution function, $f(x,a,m,b)$, given by:

$$
\begin{aligned}
f(x, a, m, b) &= [2/(b - a)][(x - a)/(m - a)] \quad \text{for } a \leq x \leq m, \\
&= [2/(b - a)][(b - x)/(b - m)] \quad \text{for } m \leq x \leq b, \tag{3.8} \\
&= 0.0 \quad \text{for other } x.
\end{aligned}
$$

Figure 3.3 shows the Excel calculations to formulate and solve Eq. 3.8. In this example, we applied the Excel solver repeatedly with different starting points to minimize the SSE in cell L13 by changing the parameter estimates in cells H4:K4. The solution derived was $a =$ any number above 3.1 and below 5.3, $m = 5.3$, and $c = 15.26$.

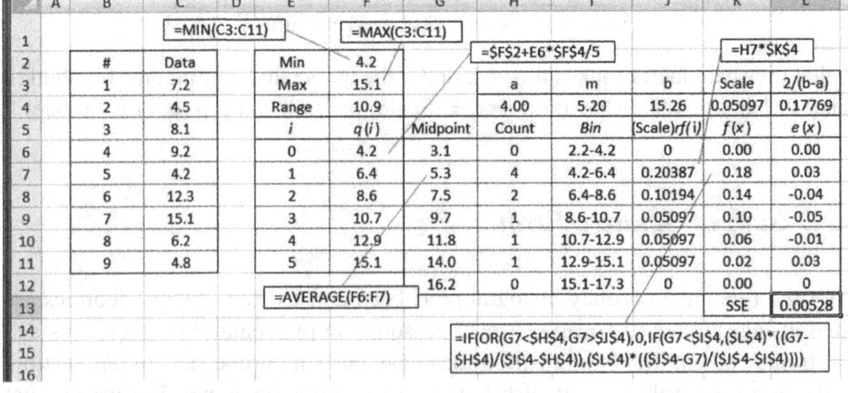

	A	B	C	D	E	F	G	H	I	J	K	L
1			=MIN(C3:C11)			=MAX(C3:C11)						
2		#	Data		Min	4.2		=$F$2+E6*$F$4/5			=H7*$K$4	
3		1	7.2		Max	15.1		a	m	b	Scale	2/(b-a)
4		2	4.5		Range	10.9		4.00	5.20	15.26	0.05097	0.17769
5		3	8.1		i	q(i)	Midpoint	Count	Bin	[Scale]rf(i)	f(x)	e(x)
6		4	9.2		0	4.2	3.1	0	2.2-4.2	0	0.00	0.00
7		5	4.2		1	6.4	5.3	4	4.2-6.4	0.20387	0.18	0.03
8		6	12.3		2	8.6	7.5	2	6.4-8.6	0.10194	0.14	-0.04
9		7	15.1		3	10.7	9.7	1	8.6-10.7	0.05097	0.10	-0.05
10		8	6.2		4	12.9	11.8	1	10.7-12.9	0.05097	0.06	-0.01
11		9	4.8		5	15.1	14.0	1	12.9-15.1	0.05097	0.02	0.03
12							16.2	0	15.1-17.3	0	0.00	0
13						=AVERAGE(F6:F7)					SSE	0.00528
14							=IF(OR(G7<$H$4,G7>$J$4),0,IF(G7<$I$4,($L$4)*((G7-				
15							$H$4)/($I$4-$H$4)),($L$4)*((J$4-G7)/($J$4-$I$4))))				
16												

Fig. 3.3 Fitting the triangular distribution to the election example data

This solution is not entirely satisfactory since we have data outside the derived range, i.e., below 5.29 and above 15.26. This likely explains why the Rockwell® Input Analyzer constrains a to equal an integer a round number below the lowest data point. Similarly b is apparently constrained to a round number above the highest data point. Then, solving m only, one derives approximately $m = 5.3$. The SSE for this solution is 0.005 (which is inexplicably different than the number from the Input Analyzer).

Note also that some distributions of interest might be characterized by important correlations between random variables. For these cases methods based on histograms only are generally insufficient by themselves. Histograms would only provide information about the marginal distributions of each random variable, i.e., the distribution of that random variable with others "averaged" out.

Procedures designed to fit the joint distributions of correlated random variables are often based on summary statistics. Active research on such cases can be found on the Winter Simulation Conference (WSC) website (http://www.wintersim.org) under the areas of input analysis, finance, and risk analysis. With these recent methods, it is becoming possible to fit and generate many correlated sets of random variables simultaneously from the same stream of pseudo random $U[0,1]$ numbers.

3.3 The Kolmogorov–Smirnov Test

Using histograms and the sum of squared error (SSE) criterion invariably involves some subjectivity. Arbitrariness enters with respect to the location of the bin endpoints. Also, arbitrary constraints can be added to the fitting optimization in Eq. 3.7 to make the result more subjectively pleasing. Next, we describe a hypothesis testing procedure which can rule out distributions with little or no subjectivity. This procedure is based on the empirical cumulative distribution function associated with a data set and the Kolmogorov–Smirnov statistic and hypothesis test.

The rigor of the KS method is perhaps misleading. This follows because, in general, no well-known or "famous" distribution is a perfect fit for any process of interest. With sufficient data, we could theoretically rule out all of the distribution that we might fit, e.g., the triangular, the uniform, the normal, and others. Related considerations motivate the application of empirical distributions described in Sect. 3.4.

3.3.1 Constructing Cumulative Empirical Distributions

The KS method is based on the so-called cumulative empirical distribution function, $F_n(x)$. For a data set, $X_1, X_2, \ldots, X_n$, the empirical cumulative distribution function is:

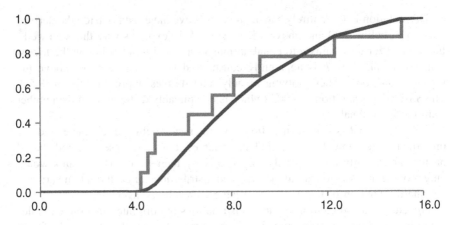

Fig. 3.4 The cumulative empirical distribution function for the election data

$$F_n(x) = (\text{the number of samples} \le x)/n. \tag{3.9}$$

Therefore, like other cumulative distribution functions it starts at 0.0 and advances to 1.0. At each data point, the function steps upward.

To create a plot of the empirical cumulative distribution function, apply the following steps:

Step 1. Sort the data from smallest to largest: $\{X_1,...,X_n\} \rightarrow \{S_1,...,S_n\}$.
Step 2. Repeat each sorted data point: $\{S_1, S_1,...,S_n, S_n\}$.
Step 3. Create sequence: $\{0.0, F_n(S_1), F_n(S_1), F_n(S_2),..., F_n(S_{n-1}), F_n(S_n)\}$.
Step 4. Scatter-line plot the sequence from Step 2 versus Step 3.

For example, consider again the $n = 9$ data points: 7.2, 4.5, 8.1, 9.2, 4.2, 12.3, 15.1, 6.2, and 4.8. Then, we have $S_1 = 4.2,..., S_n = 15.1$. Also, the two sequences are $\{4.2, 4.2,..., 15.1, 15.1\}$ and $\{0.0, 0.111, 0.111, 0.222,..., 1.0\}$. Then, the empirical cumulative distribution function is shown in Fig. 3.4. The empirical distribution function, $F_9(x)$, is the step function in the figure. The smooth function is the cumulative distribution function for the fitted distribution, $F(x, \beta_{est})$.

3.3.2 The Kolmogorov–Smirnov Test

The Kolmogorov–Smirnov (KS) test is based on the test statistic D_n given by:

$$D_n = \text{Maximum}\{F_n(x) - F(x, \beta_{est})\} \tag{3.10}$$

where the maximization is taken over the values x that the random variable might assume. For example, with the $n = 9$ data points: 7.2, 4.5, 8.1, 9.2, 4.2, 12.3, 15.1, 6.2, and 4.8 and the fitted distribution $a = 4.0$, $m = 5.2$, and $b = 16.0$, the KS statistic is $D_9 = 0.289$. This maximum value occurs at $x = 4.8$.

For completeness, the general formula for the cumulative triangular distribution function, $F(x)$, is:

$$F(x, a, m, b) = \left[(x - a)^2 \right] / [(b - a)(m - a)] \quad \text{for } a \leq x \leq m$$
$$= 1 - \left[(b - x)^2 \right] / [(b - a)(b - x)] \quad \text{for } m \leq x \leq b. \tag{3.11}$$

This is the cumulative (the smooth curve) plotted in Fig. 3.4 together with the cumulative empirical distribution function, $F_9(x)$.

Does $D_9 = 0.289$ indicate a significantly poor fit? In KS hypothesis testing, the test statistic is compared with a critical value. If the statistic is larger than the critical value, then we reject the hypothesis that the data came from the fitted distribution. The event that $D_n >$ critical KS is equivalent to the event that the p value is <0.05. In our example, standard software can tell us that the p value is 0.15. Therefore, we fail to find significance. Essentially, we do not have sufficient data to rule out the fit distribution which is triangular, i.e., the data might conceivably have been generated from the triangular distribution with parameters $a = 4.0$, $m = 5.2$, and $c = 16.0$.

Deriving the critical values and the p values is a laborious process accomplished commonly by software. Yet, in principle the method for their derivation is clear. One can simulate 10,000 sets of n data from the fit distribution using its inverse cumulative. In every case we can derive a simulated value for D_n. We can take the 500th largest value as our 95% critical value. If the actual KS is larger than this value it likely did not occur by chance. In that case, the distribution is a significantly poor fit.

3.4 Empirical Distributions

It is perhaps true that, with sufficient data, any distribution can be ruled out using KS hypothesis testing. Generally, as the amount of data goes to infinity even slight departures from fit distributions are detectable. In cases in which all the famous distributions have been ruled out, it is entirely natural to simply sample from the data instead of simulating using fitted distributions. Sampling from the data with replacement is equivalent to applying a fit distribution which is the cumulative empirical distribution in Eq. 3.9.

As an example consider the relative frequency histogram from 719 actual voting times from the 2006 gubernatorial election in Franklin, Ohio in Fig. 3.5. These times were read from a single machine and do not reflect the time needed for the poll worker to prepare the machine and prepare the voter. Even with 719 data points, the KS test can rule out the triangular (p value <0.005), the normal (p value <0.005), and all other distributions in the Rockwell® Input Analyzer®. This motivated the choice to apply empirical distribution in our election simulations for the voting or service times.

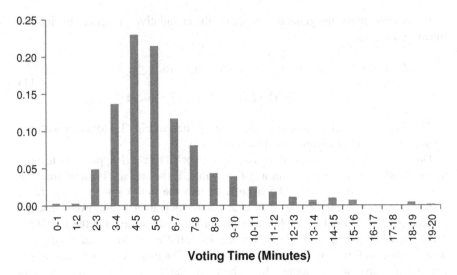

Fig. 3.5 Relative frequency histogram of actual DRE utilization times from 2006

Because these times represent one of the initial experiences of the community with handicapped compliant machines, they were a "wake-up" call for election officials. The new machines had roughly double the service times of previous machines. Therefore, it was perhaps true that nearly double the number of machines would be needed for the 2008 presidential election as compared with the 2004 presidential election. Fortunately, early voting and paper ballots on Election Day headed off what simulation predicted would have been extremely long election lines.

Even in cases with limited data, empirical distributions have appeal and appear to be in common usage. Yet, in these cases it may be desirable to insert hypothetical or "made-up" data in with the real data. Such an approach is in keeping with the subjective view of probability theory described in Chap. 2. Some common sense considerations relating to made-up data include:

- Using made-up data should be accompanied with a full *disclosure* and clarification of what is real and what is assumed,
- *Scaling* of the data may be needed to adjust to new cases, e.g., additional work planned in the service operations,
- *Sensitivity analyses* should be included in related projects to evaluate the extent to which the final conclusions of the simulation study depend on data made-up during the input analysis stage, and
- *Proportionality* should be considered such that real data can be duplicated to permit the made-up data to occur in the data set with the appropriate frequencies. Since we will be sampling with replacement, it is important that each point has at least roughly the appropriate probability of being selected.

Despite the challenges of disclosure and sensitivity analysis, inserting hypothetical data and building empirical distributions can, in some cases, be part of a conservative analysis. For example, it has been reported that several operations researchers mitigated portfolio losses in the recent downturn using empirical distributions with made-up data corresponding to problems not encountered in previous years.

In our election systems projects, we use empirical distributions with data scaled to address the variable ballot lengths in different locations. The data we have is for specific ballot lengths from the 2006 gubernatorial election for 6 and 8 issues. When planning for the 2008 presidential election, we knew that the ballots would be as much as 2.1 times longer. As a result, we scaled our data linearly for each different location to model voting service times. Since all fitted distributions were ruled out, this seemed to us to be the most defensible approach given our budget limitations. Our mock election results and, later, observations of the actual voting times validated this approach.

3.5 Summary Example

In this section, an example is given that illustrates the cycle of data collection, distribution fitting and analysis, and use of the fitted distributions for simulation. In the example, we assume that 0.82, 0.14, 0.56, 0.31, and 0.90 are pseudo random $U[0,1]$ numbers. Also, we assume the following hypothetical times (in minutes) from observing bus arrivals starting at 10 am on 11 different days: 0.5, 0.2, 1.1, 0.8, 0.9, 1.4, 1.9, 0.8, 2.0, 0.0, and 0.8. Questions to be answered include:

1. Is this sample size advisable and what are the related issues?
2. What is the continuous distribution associated with a relative frequency histogram?
3. Give a best fit distribution based on the SSE.
4. How could the distribution that was fitted be ruled out hypothetically using a KS test?
5. Generate some simulated times using the best fit distribution.

Answers follow. (1) Generally, having fewer than 20 samples is not desirable since we have limited ability to estimate the sample variance. Also, we cannot eliminate distributions using KS testing. Certainly, we have too few samples if accounting for any type of rare event is needed for predicting system performance accurately.

(2) Creating a relative frequency histogram and the associated continuous distribution aids in distribution fitting. Figure 3.6 shows the associated continuous distribution. (3) Next, we determine the best-fit uniform distribution that minimizes the SSE in Eq. 3.7. The choice of the distribution is somewhat arbitrary but the shape of the uniform distribution seems reasonable. The continuous distribution associated with the relative frequency histogram is shown in Fig. 3.6. The best

Fig. 3.6 The continuous
distribution associated with
the relative frequency
histogram and the best fit
uniform distribution

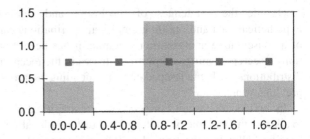

Fig. 3.7 Cumulative
empirical distribution and
cumulative best fit
distribution

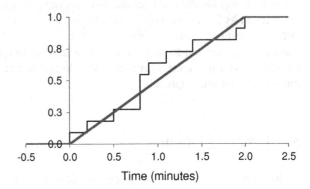

fit parameters are (apparently) $a = 0.0$ and $b = 2.0$ and the sum of squared errors
is 1.1. The parameters were obtained by manually trying many combinations of
fitted a and b values and checked using the Excel solver.

(4) The empirical cumulative distribution is shown in Fig. 3.7 (the step
function). The smooth line shows the cumulative distribution, $F(x)$, for the best
fitted uniform distribution. The biggest difference is approximately 0.19 (vertical
absolute value) which occurs at 0.9 min. In formal KS-testing, this value is
compared with a 95th percentile value derived from simulating KS statistics using
the assumed distribution. With only 11 values, it is unlikely that the test statistic
will be beyond the critical value regardless of which distribution is selected.

(5) Using Eq. 2.18 we have $F^{-1}(u) = 2u$. Plugging in pseudorandom numbers,
we generate the five simulated times: 1.6, 0.28, 1.1, 0.6, and 1.8 min. These
simulated numbers can be used with an event controller to estimate the expected
waiting times or other emergent system properties.

3.6 Problems

1. What is an empirical cumulative distribution function?
2. What is a relative frequency histogram?
3. Why is picking a distribution based on the SSE and relative frequency
   histograms somewhat arbitrary?

4. Why is testing based on the KS statistic a rigorous and repeatable process, i.e., what (if any) arbitrariness exists in the process of calculating KS statistics?
5. Use the following to construct a relative frequency histogram based on the 20 data points and estimate by eye the sample mean and standard deviation.

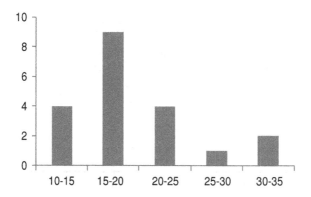

6. In the preceding problem, how many bins would Sturges' rule recommend?
7. Use your eye to best fit a distribution to the histogram in problem 5. Estimate its SSE based on your relative frequency histogram.
8. Consider the following data: 2, 11, 9, 9, 12, 15, 12, 3, 10, and 9. Construct a relative frequency histogram, fit a distribution, and estimate the SSE using a spreadsheet and showing all work.
9. In the following relative frequency histogram, what is the height of the missing bar? Also, what role do relative frequency histograms play in our distribution selection?

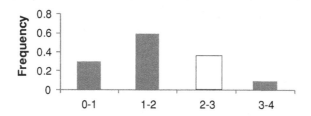

10. In the preceding problem, estimate the SSE for a best fit normal distribution.
11. What is the following plot and what could its role be in input analysis?

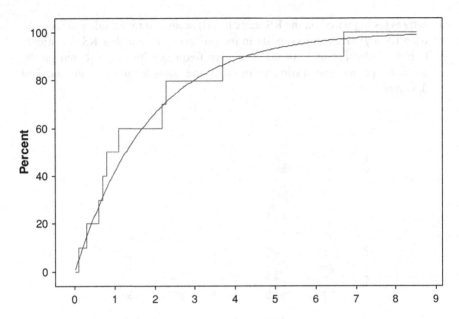

12. Estimate the KS statistic in the preceding problem. Do you guess that the KS test would reject (p value <0.05) in that problem? Explain briefly.
13. Consider the data which represent hypothetical voting times in minutes: 5, 10, 5, 22, 6, 12, 14, 8, 7, and 4. Construct a relative frequency histogram of this data.
14. Consider the data which represent hypothetical voting times in minutes: 4, 22, 5, 5, 9, 12, 14, 8, 7, and 4. Construct a relative frequency histogram of this data.
15. List one distribution that might be ruled out using KS testing for the dataset in problem 13.
16. Determine D_{10} for the data in problem 13 and a reasonably appropriate triangular distribution function.

Chapter 4
Simulating Waiting Times

This chapter describes the standard process for performing discrete event simulations to estimate expected waiting times. In doing so, it describes event-based "controllers" that generate chronologies differentiating discrete event simulation from other types of statistical simulations.

Generic waiting systems have arrival processes, service processes, and departure processes as indicated in Fig. 4.1. In general, the queues or waiting lines form in front of the service processes. Simply applying the terminology of arrival and service processes can sometimes help stakeholders think more constructively about systems improvement. For example, in election systems we have observed that suspicion of intentional prejudicial action tends to be reduced once the more "scientific" or "clinical" framing of the problem in Fig. 4.1 is introduced.

The chapter begins in Sect. 4.1 by describing probably the most common arrival process, which is characterized by exponentially distributed "interarrival" times. Interarrival times, as the name implies, are the times between arrivals. Exponential interarrivals relate to the exponential distribution described in Chap. 2. Exponential interarrivals are associated with so much variation that they are generally irrelevant to many manufacturing or service operations. Manufacturing processes are generally more tightly controlled and less variable. Yet, exponential interarrivals are common enough that they have been dubbed with a well-known name as "Poisson" arrivals. This name derives from the fact that, with exponential interarrivals, the number that accumulates in any given time period is a Poisson random variable. Poisson arrivals are generally relevant when the entities arriving are uncoordinated such as customers entering a store individually.

In Sect. 4.2, we describe the general operation of an event-driven simulation controller. Such a controller is the key element of virtually any discrete event simulation software. The controller effectively observes all of the possible events that could happen, picks the closest in time, and moves the clock forward. The resulting series of event times is "aggregated" to estimate the means of interest.

T. T. Allen, *Introduction to Discrete Event Simulation and Agent-based Modeling*,
DOI: 10.1007/978-0-85729-139-4_4, © Springer-Verlag London Limited 2011

Fig. 4.1 Generic waiting system showing arrivals, waiting in queue, service, and departure

Next, in Sect. 4.3, by inspection of the central limit theorem (CLT) and the conditions that underlie confidence intervals, we show that multiple replicates of the entire time period are generally helpful for defensible estimation. Further, motivated by the CLT, we describe in Sect. 4.4 the process of batching replicates in groups whose sample average might reasonably be expected to be approximately normally distributed. Such a process constitutes probably the leading way to produce defensible Monte Carlo estimates of expected values and their half widths.

4.1 Exponential Interarrivals or Poisson Processes

The exponential distribution function, $f(x)$, cumulative distribution function, $F(x)$, and cumulative inverse distribution function, $F^{-1}(u)$ are given by:

$$f(x) = (\lambda)[\exp(-\lambda x)], \tag{4.1}$$

$$F(x) = 1 - \exp(-\lambda x), \quad \text{and} \tag{4.2}$$

$$F^{-1}(u) = -\ln(1 - u)/(\lambda) = -(\text{mean})[\ln(1 - u)] \tag{4.3}$$

where "exp()" means take 2.7182818 to the power of the number in the parentheses and "ln()" is the natural logarithm, i.e., the logarithm with base "e" equal to 2.7182818. Therefore, to generate pseudorandom exponentially distributed random numbers with mean equal to the reciprocal of λ, we plug pseudorandom $U[0, 1]$ numbers into Eq. 4.3. For the "two parameter exponential" distribution, simply replace x with $x - T$ in Eqs. 4.1 and 4.2 and add T in (4.3).

For example, consider a Poisson process characterized by an assumed mean interarrival time of 5 min. Therefore, $\lambda = 0.2$ arrivals per min. Also, if we have a pseudorandom $U[0, 1]$ number or "deviate" given by $u = 0.5636$ then we have a pseudorandom number (PRN) from the exponential distribution given by:

$$F^{-1}(0.5636) = -(5.0\,\text{min})[\ln(1 - 0.5636)] = 4.1\,\text{min}. \tag{4.4}$$

The exponential distribution has some strange properties which can make us concerned when we use it to describe our beliefs. For example, it is "memoryless" which relates to the process of waiting for arrivals. If one has already waited Z minutes for an exponential interarrival characterized by λ, the expected wait is still $(1/\lambda)$. In plain English, the memoryless property means that the distribution gives us no credit for the time we already waited. For example, if $\lambda = 0.2$ arrivals per min and one has already waited for 5.0 min, amazingly the expected waiting time is still 5.0 min.

The memoryless property of the exponential makes it an unreasonable choice for most arrivals on a schedule or arrival processes that have many kinds of constraints. For example, if a certain number of people need to arrive in a certain short time period, the exponential distribution might not be an appropriate choice for interarrival times. Still, the life expectancy of many electronic components and certain subatomic particles follow the memoryless exponential distribution very closely. Also, partly because of its simplicity, Poisson arrivals are probably the world's most common arrival process for modeling uncoordinated arrivals over short periods.

Over longer periods, it is often reasonable to assume that the parameter λ governing exponential interarrivals changes over time. For example, for the first hour of the day we might have $\lambda = 0.2$ per h after which the frequency of arrivals might pick up. Then, we could have $\lambda = 0.3$ per h for the second hour. The phrase "nonhomogeneous Poisson process" refers to the assumptions that, at any given time, interarrivals are exponentially distributed according to one parameter value, yet, over time that parameter changes. Nonhomogeneous Poisson processes are one example of "nonstationary" arrival processes which change over time. Nonhomogeneous Poisson processes are probably the most common assumption about arrivals in discrete event simulation models despite their highly variable nature. Additional details about nonhomogeneous Poisson processes are described in Chap. 10 related to the "thinning" method.

4.2 Discrete Event Simulation Controllers

In Chap. 2, we showed how it is possible to apply simulation to derive an estimate of an expected value or mean. The example that we focused on was simple in the following sense. The experiences of the simulated individuals were unrelated, i.e., if one took a long time to vote using the direct recording equipment (DRE) machine, this had no effect on other individuals. In this section, however, we focus on simulating waiting times. Waiting times necessarily relate to interactions between individuals. Specifically, one individual is using the resource for which one or more other individuals are waiting or queuing. As noted previously, this "event-chronology-based" interaction between entities is what differentiates discrete event simulation from other forms of Monte Carlo.

Discrete event simulation systems are driven by events in a chronology or time listing of events and the byproducts from that chronology. Other alternatives have advantages compared with event based simulation including the so-called "three phase" method (Pidd 2004). Here, we focus on discrete event simulation because it is probably the simplest and clearly the most widely applied in practice.

Code for discrete event simulation invariably includes an event controller. The controller is a computer agent with two main tasks. First, it identifies the list of events that could possibly happen next, if any. If there are no possible events, e.g., because the replication is over, the controller stops. Second, after a list of possible

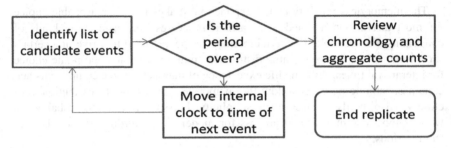

Fig. 4.2 A generic controller for a single replicate of discrete event simulation model

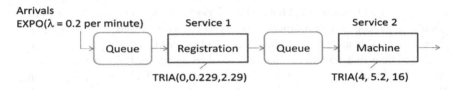

Fig. 4.3 Simplified voting system with waiting lines

events is compiled, the controller advances the internal clock to the time of the first event on the list of possible events. Then, iteration continues as indicated in Fig. 4.2. The aggregation of numbers refers to the building averages and other statistics such confidence intervals. These numbers are derived from the chronology or list of events and times generated by the controller and stored. Here, we focus on numbers which are average waiting times, yet simulations can also generate statistics with scope ranging from profits to machine down times to the number killed in warfare. Having generated and (potentially) stored a complete account of simulated events, the simulation can support estimation of a wide variety of expected values.

As an example, consider the simplification of the voting systems in Fig. 4.3. The work flow in the figure is a simplification because it does not include automobile parking issues, the precinct having more than a single machine, and the details of the departure processes. Also, in our analysis we assume for simplicity that the voting day is only 20 min long, i.e., only arrivals in the first 20 min are allowed entry. In real voting systems, the minimum number of machines is generally greater than or equal to three at each polling location to allow for machine breakdowns and other contingencies. Voting is generally permitted in the United States for 13 h or more and, increasingly in many jurisdictions; it is permitted for many days prior to "Election Day" at certain publicized locations.

Figure 4.4 shows the event controller applied to the simplified election system example (Fig. 4.3). The $U[0, 1]$ PRN stream is shown on the left-hand side of Fig. 4.4. This stream derives from the Data Analysis → Random Number Generation component of Microsoft® Excel (in the Analysis ToolPak). As noted previously, linear congruential generators (LCGs) are generally usable only for

U[0,1]	PRN. Time	Type	Event PRN.	P1 arr. (1)	P1 reg. (2)	P2 arr. (3)	P2 reg. (4)	P3 arr. (5)	P3 reg. (6)	P1 dep. (7)	P4 arr. (8)	P4 reg. (9)	P2 dep. (10)	P3 dep. (11)	P4 dep. (12)
0.0013	P1 interarrival	EXPO(5)	0.01	0.01											
0.5636	P2 interarrival	EXPO(5)	4.15			4.15	4.15								
0.1933	P1 reg.	TRIA(0,0.229,2.29)	0.34		0.35										
0.8087	P1 voting	TRIA(4,5.2,16)	11.02				11.37	11.37	11.37	11.37	11.37				
0.5850	P3 interarrival	EXPO(5)	4.40					8.55	8.55						
0.4799	P2 reg.	TRIA(0,0.229,2.29)	0.72				4.88								
0.3503	P3 reg.	TRIA(0,0.229,2.29)	0.54						9.09						
0.8960	P4 interarrival	EXPO(5)	11.32							19.86	19.86	19.86			
0.8228	P2 voting	TRIA(4,5.2,16)	11.21									22.57	22.57	22.57	
0.7466	P5 interarrival	EXPO(5)	6.86									Late			
0.1741	P4 reg.	TRIA(0,0.229,2.29)	0.32									20.18			
0.8589	P3 voting	TRIA(4,5.2,16)	11.72											34.30	
0.7105	P4 voting	TRIA(4,5.2,16)	9.87												44.17

Fig. 4.4 A single replication showing the controller operation in the election systems example

Table 4.1 Aggregate/summary statistics from the first replicate of the voting systems simulation

Voter	Registration queue time (min)	Vote queue time (min)	Sum
Person 1 (P1)	0.00	0.00	0.00
Person 1 (P2)	0.00	6.49	6.49
Person 1 (P3)	0.00	13.49	13.49
Person 1 (P4)	0.00	14.12	14.12
		Average (min)	8.52

instruction purposes, i.e., to help students comprehend the concept of random seeds and random number streams.

In this system, the only candidate initial event is the arrival of the first voter, i.e., "person 1" (P1). Clearly, neither the arrival of person 2 (P2) nor the conclusion of the first voter's registration can precede this. Therefore, the clock moves forward to 0.01 min when the P1 arrives. Next, the arrival of P2 and the end of the first voter's registration are candidate events. Based on numbers derived from the inverse cumulative distribution functions (F^{-1}), P1's finishing registration is the next event and the clock advances to $0.01 + 0.34 = 0.35$ min. Note that the P2 arrival is $0.01 + 4.15 = 4.15$ min because of the rounding in excel of instead of 4.16.

By random draw, P1 takes a long time while occupying the single DRE voting machine, i.e., 11.02 min. During that time P2 completes registration and person 3 arrives and finishes registration. When P1 finishes voting and departs, both P2 and P3 are waiting. After P2 has begun voting, person 4 arrives and registers. The arrival of person 5 would come after the 20 min hypothetical Election Day is over. Therefore, person 5 is turned away and does not influence the statistics. Next, the remaining people finish up voting and depart.

After the controller has generated the chronology, extracting statistics and aggregating them is possible. Often, a single number is drawn to effectively summarize the experience in the replicate. This summary or "aggregate" statistic often is associated with evaluating the system quality. For example, relating to the chronology in Fig. 4.4, Table 4.1 includes the event statistics and the aggregate statistic, which is the average waiting time. The single number that summarizes quality from this replicate is the average waiting time of 8.53 min.

It is tempting to use the sequence, 0.00, 6.49, 13.49, and 14.12 min to create a confidence interval for the mean or expected value of the waiting time. However, by studying closely the definitions of statistical independence, identically distributed, and normally distributed in the next section, we will see that this is not wise. The resulting interval would have a much <95% chance of containing the true mean.

4.3 IID Normally Distributed

Next, we review the definitions of statistical independence, identically distributed, and normally distributed. These definitions and their implications motivate the procedure in the next section which involves multiple replicates and a technique known as "batching for normality". Using these techniques, the resulting confidence intervals for the means or expected values of interest will be fully defensible. The chance that 95% confidence intervals actually contain the mean will be actually 95% to a good approximation.

Technically, two random variables, X_1 and X_2, are independent if and only if (iff) their joint distribution, $f(x_1, x_2)$ factorizes. If $f(x_1)$ and $f(x_2)$ are the marginal distributions, we have independence iff:

$$f(x_1, x_0) = f(x_1)f(x_2). \qquad (4.5)$$

In words, independence means that if you tell me the actual value of X_1, my beliefs about X_2 are unchanged. This typically occurs if there is no causal link between the systems that generate X_1 and X_2. Note that one way to demonstrate that two random variables cannot reasonably assumed be independent is to observe a significant sample correlation in their historical data streams, i.e., when one value was higher (lower) than usual for X_1, the corresponding X_2 was typically higher (lower) than usual.

More importantly, the following is a fact stated without proof: if we repeat simulations using independently distributed random number inputs, the derived aggregate outputs will also be independent. The intuitive explanation for this fact is that there is no reason to believe that, if the first replicate output is low, then the second replicate output will be low or high. Further, consider the simulation method based on applying an industrial quality PRN $U[0, 1]$ stream pictured in Fig. 4.5. The construction method and the abovementioned fact together explain why we are generally comfortable assuming that simulation outputs constructed this way, $X_1, X_2, X_3, \ldots$ are independently distributed. In our examples, this will mean that the average waiting times from different replicates are reasonably assumed to be independently distributed.

Technically, two random variables, X_1 and X_2, are identically distributed if and only if they have the same marginal distribution function, $f(x)$. Therefore, if $f_1(x_1)$ is the assumed marginal for X_1, and $f_2(x_2)$ is the assumed marginal for X_2, then the marginal distributions satisfy:

Fig. 4.5 How controllers create approximately independent outputs

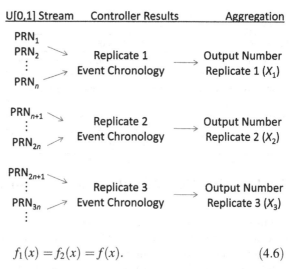

U[0,1] Stream	Controller Results	Aggregation
PRN$_1$		
PRN$_2$	Replicate 1	Output Number
⋮	Event Chronology	Replicate 1 (X_1)
PRN$_n$		
PRN$_{n+1}$	Replicate 2	Output Number
⋮	Event Chronology	Replicate 2 (X_2)
PRN$_{2n}$		
PRN$_{2n+1}$	Replicate 3	Output Number
⋮	Event Chronology	Replicate 3 (X_3)
PRN$_{3n}$		
⋮		

$$f_1(x) = f_2(x) = f(x). \tag{4.6}$$

In words, X_1 and X_2 are identically distributed random variables if they "come from" the same distribution. This brings us to the next fact again related to Fig. 4.5. If we replace the PRNs with true random numbers, then $X_1, X_2, X_3,\ldots$ will be identically distributed random numbers. If we use PRNs, then it is generally reasonable to treat the outputs as if they were identically distributed random numbers. Intuitively, this follows because the controller applied the same rules to generate the event chronology with only the input PRNs changing.

A random variable, X, is normally distributed if and only if its distribution function has the following form:

$$f(x) = [(6.283185307)(\sigma^2)]^{-1/2} \exp[(x - \mu)/(2\sigma^2)] \tag{4.7}$$

where μ and σ are assumed parameters with $\sigma > 0$. One pleasing property of the normal distribution is that its mean and standard deviation directly correspond to its adjustable parameters, μ and σ, respectively. This property is pleasing because it implies that we do not need to apply calculus or simulation to estimate the mean from the distribution function. In Chap. 3, we describe how data can be competently used to identify distribution functions that are reasonable choices. Also, hypothesis testing to rule out distributions is described.

In the context of our election systems example, it is obvious that the waiting time numbers are not even approximately independent nor identically distributed (IID). Table 4.2 describes reasons related to each issue.

Technically, one violation of either independence, identically distributed, or normally distributed could be sufficient to render the standard confidence intervals unlikely to reliably contain the true means. For the numbers generated within a single replication, we generally have violations of all three assumptions. Fortunately, in the next section we describe a way to derive pseudorandom outputs from simulations that can defensibly be assumed to be IID normally distributed.

Table 4.2 Why variables within a single replicate are not approximately IID normally distributed

Property	Reason why not
Independent	If one simulated person waits a long time that will likely cause the next simulated person to wait a long time
Identically distributed	The first simulated person never waits, i.e., has zero mean waiting time. Others have nonzero mean waiting times
Normally distributed	Waiting times cannot be less than zero. The normal distribution has technically zero weight on any point. Yet, we believe that there is a nonzero chance for a waiting time equal to 0.00 min

4.4 Batching Complete Replicates for Normality

In this section, we review the CLT. Then, we describe a "batching for normality" procedure that generates perhaps the most defensible outputs from discrete event simulations possible. By applying high-quality PRNs from the $U[0, 1]$ distribution, independent replication, and batching for normality, the generated outputs are independent identically distributed (IID) and normally distributed to a good approximation for virtually any type of discrete event simulation.

4.4.1 The Central Limit Theorem

Assume that $X_1, X_2,...,X_n$ are independent, identically distributed (IID) according to any distribution with any mean, μ, and finite standard deviation, σ_0, then the CLT give us:

$$\text{Limit}\{n \to \infty\}\ X_{\text{bar},n} \equiv (X_1 + X_2 + \cdots + X_n)/n \sim N[\mu, \sigma_0/\text{sqrt}(n)] \qquad (4.8)$$

where "$N[]$" implies normally distributed as defined by Eq. 4.8. In words, the CLT says that the sample average, $X_{\text{bar},n}$, of n IID outputs from simulation replicates is approximately normally distributed for a sufficient number of replicates (n large enough). In Chap. 3, we describe technology for evaluating whether n is large enough in any specific case. Yet, the key fact is that there is almost always a large enough n such that normally distributed output averages are possible.

The central limit theory essentially explains how and why discrete event simulation works. Through the simulation controller and aggregation we derive IID $X_1, X_2,...,X_n$. With a large enough number of replicates, n, our Monte Carlo estimate for the mean, $X_{\text{bar},n}$ converges to the true mean or expected value that we are trying to estimate. Further, the errors become approximately normally distributed with standard deviations declining proportional to $\sigma_0/\text{sqrt}(n)$ as n increases. Because $X_{\text{bar},n}$ converges to the true mean, we say that Monte Carlo gives "unbiased" estimates. This contrasts with certain alternative procedures which give biased estimates as described in Chap. 8. Deriving a slightly wrong number quickly and reproducibly can be of interest in certain situations.

Fig. 4.6 An example discrete probability density function

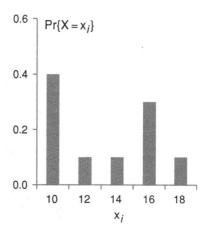

As an example, consider the probability mass function indicated in Fig. 4.6. Obviously, this distribution is not the normal distribution. It is not even the distribution of a continuous random variable. The distribution is for discrete random variables, $X_1, X_2,...,X_n$. Yet, clearly these variables have finite standard deviation, σ_0. Using the definition of the mean or expected value, μ, in Eq. 2.2, the mean is 13.2. Also, using the definition that the standard deviation equals the square root of $E[(X - \mu)^2]$, we have σ_0 equal to 6.57.

Next, consider the distribution of the sample averages, $X_{\text{bar},n}$, of batches of X_1, $X_2, ..., X_n$ generated from the distribution in Fig. 4.6. As a practical matter, we can generate these using the "Data Analysis" $\rightarrow$ "Random Number Generation" facility in Microsoft® Excel and selecting "Discrete" among the types of distributions. Doing this, we can use "Number of Variables" equal to 10 and "Number of Random Numbers" equal to 1,000. This will generate 1,000 rows by 10 columns of PRNs. Averaging the first $n = 5$ or $n = 10$ in each row gives sample averages, $X_{\text{bar},n}$. From the CLT, we suspect that the averages based on $n = 10$ numbers will appear normally distributed to a better approximation. The histograms overlaid in Fig. 4.7 shows that the sample averages are becoming more normally distributed, i.e., more bell shaped as n increases.

4.4.2 Batching for Normality

The CLT motivates a strategy for developing the approximately normally distributed PRNs needed for proper confidence intervals. We reviewed the procedure for creating defensible intervals in Sect. 2.2. By batching outputs from independent replicates, the batch averages become approximately normally distributed. The approximation improves as the sample size, n, increases.

Figure 4.8 shows an example of batching for normality with a batch size equal to $n = 3$. The implication from studying the figure is that the outputs derive from a

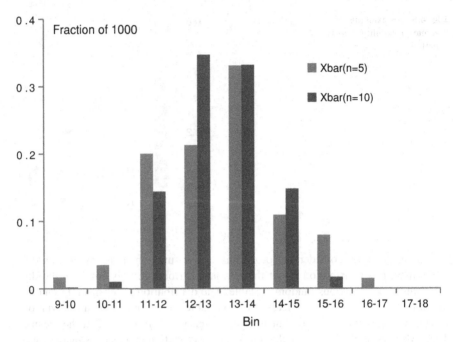

Fig. 4.7 Histograms of sample averages with $n = 5$ and $n = 10$ in each batch

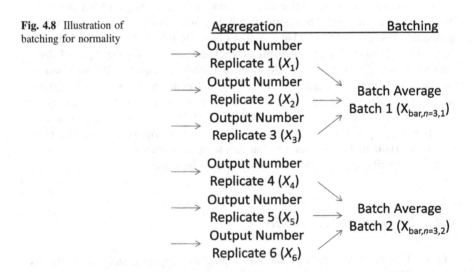

Fig. 4.8 Illustration of batching for normality

standard event controller and aggregation process. Whether n needs to be increased can be tested using methods in Chap. 3, however, the central limit theory indicates that the larger the n, the more likely normality will be achieved.

Note that the Monte Carlo estimate for the mean is unaffected by batching. It is still the average of the batch averages. This is the same as the average of all the replication outputs. What is affected is the quality of the estimated standard deviation used in constructing a confidence interval for the mean. By batching, the averages are more normally distributed and the half width is more likely to be a reliable estimate of uncertainty.

4.5 Other Arrival Processes

In this chapter, the Poisson arrival process has been introduced and defined in terms of exponentially distributed interarrival times. Poisson processes are generally considered relevant for modeling uncoordinated arrivals such as patrons arriving at restaurants or voters arriving to polling locations on Election Day. Still, many manufacturers use so-called "pull systems" in which units are made only when they are ordered. Pull systems are one component of the Toyota production systems or lean production. It can be reasonable to model pull systems using Poisson processes because the orders do arrive in an uncoordinated fashion, e.g., at the request of the dealerships.

At the same time, scheduled arrivals and scheduled builds might be a safer choice in general. It is not uncommon to use Visual Basic or other programming languages to enter data about arrivals into models developed using standard software such as ARENA, AutoMod, ExpertFit, GPSS/H, ModSim, or WITNESS. For a systematic review of related simulation software (see Swain 2007). In this way, the arrival process for the projected future can be modeled directly using historical data.

The term "push system" refers to operations which make units following long term forecasts for demand. Such systems might be expected to leave machines idle more rarely, i.e., "starve" their servers. In these systems, units are produced to keep machines utilized and in the belief that what is made will be purchased. Push systems can be modeled using a high rate Poisson order arrival process. However, such an approach can make it difficult to limit the number of PRNs used in each replicate. Conserving PRNs helps speed up simulations and facilitates comparisons of alternatives because alternatives can essentially be evaluated more easily using the same set of random numbers and situations. We will have more to say about speeding up simulations in Chap. 8.

4.6 Model Verification and Validation

It is standard to define "verification" as checking that the simulation model does as intended, i.e., there are no "bugs" or simple mistakes in its complications or procedures (Sargent 2005). Validation relates to the ability of a verified model to reproduce the outputs of the system being modeled accurately. Validation therefore relates to the appropriateness of the model entity flows, interactions, and distributional assumptions. Validating models is generally more complicated but both verification and validation are important.

Both verification and validation can be conducted simultaneously using the following activities as described by Sargent (2005) and Czeck et al. (2007):

- *Design of experiments output analysis* (Chap. 5) this involves structured experimentation to determine how input changes affect average outputs. Inspecting the plausibility of the dependencies often permits the identification of either bugs or unrealistic assumptions.
- *Removing randomness and running test cases* often, all random variables are set to constants at mean values can facilitate evaluation of a model's logic. In some cases, there is an attempt to line up the simulated entity positions with real-world locations at multiple times using photographs or other data sources.
- *Scatter plotting or quantitative output validation* outputs from simulation replicates or averages can be graphed against real-world data from the associated systems. The extent to which predicted and actual data line up (have high sample correlation) provides one of the strongest types of confirmation possible. Also, the maximum percentage deviations can evaluate the accuracy.
- *Turing tests* subject matter experts (SMEs) are shown both simulated and real-world data. If the SMEs identify the simulated data, reasons are solicited. The derived information is then used to improve the simulation. Once the simulated data cannot be distinguished from the real by the SMEs the model is considered validated. Since the test involves fooling people about the realism of computer models, it is reminiscent of the so-called "Turing" tests in which computers attempt to convince humans that the computers are human.
- *Animated walkthroughs with team members* trust in models is generally critical for acceptance of the derived recommendations. Animation permits the stakeholders to evaluate the model and go through step-by-step replicates visually to ensure some level of realism.

Many organizations that use simulation regularly establish standards relating to their validation. For example, a dentist's office might feel that reducing waiting times is critical for its success. They might seek the most objective and thorough possible validation. As a result, they might focus on scatter plot or quantitative validation to study the "base case" simulation model. The simulated waiting time average and maximum before being taken to an examination room for prophylaxis were 2.6 and 8.0 min, respectively, for the first hygienist, and 2.7 and 7.1 min for the second hygienist. At the same time, the observed waiting time average and maximum for the dentist's consultation after prophylaxis were 2.8 and 7.3 min based on 20 measurements. The base case model matched all six of these observations to within 10%.

As a second example, a manufacturing company may be studying fairly simple material handling systems with easy access to the real-world positions of entities. As a result, the manufacturer might generally require removing randomness and running test cases based on time photographs of parts and try to exactly replicate part positions. This verification and validation might become part of all model developments.

4.7 Summary Example

In this chapter, we considered the process of creating and analyzing an event chronology based on a PRN stream. While the event chronology generates several simulated times, problems are associated with these simulated times. Most importantly, the numbers within any given chronology are not reasonably assumed to be independent and identically distributed (IID). As a result, using these numbers alone to construct confidence intervals for the expected properties generally results in intervals that likely do not include the true mean.

As an example, consider the simulation and event controller in Fig. 4.9 based on simulated voting for a hypothetical 20-min long Election Day, i.e., arrivals after 20 min are turned away. The simulation starts with pseudorandom IID uniform $a = 0$ and $b = 1$ numbers (Set A). Next, the set of immediately following possible events are enumerated and each of the original uniform PRNs is transformed into a PRN from the appropriate distribution. These numbers are then used to generate the times of the next events (Set B) and newly possible events are enumerated. A program that automatically assembles the chronology of events in this way is called an event controller. From the chronology it is possible to calculate how long each simulated entity waits. These numbers are collected in the column on the right-hand side of Fig. 4.9 (Set C). The sample average of these waiting times (Set D) is an estimate of the expected waiting time.

The example in Fig. 4.9 leads to many questions which are addressed in the chapter:

1. Which sets of numbers are reasonably assumed to be IID?
2. Which sets of numbers are approximately normally distributed?
3. Why are Set C numbers not independent and not identically distributed?
4. Suppose that we have average waiting times from six replicates (times in minutes), 8.5, 22.1, 7.1, 3.7, 8.2, and 9.2. How can we make a defensible confidence interval from these numbers?

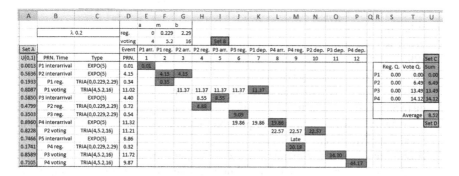

Fig. 4.9 An example illustrating a discrete event simulation and sets of random variables

Answers follow. (1) While we do not have the details about how the pseudorandom uniform numbers in Set A were generated, it is probably reasonable to assume that the Set A numbers are IID. The Set A numbers show no obvious pattern and it is easy to generate pseudorandom uniform IID [0, 1] numbers. However, the other sets do not contain numbers that can be reasonably assumed to be approximately IID. The process of constructing the chronology leaves it clear that the absolute times of events (Set B) and the waiting times (Set C) are influenced by the same simulated activities. They have, therefore, built-in causal links or dependence. Clearly, with a single number only in Set D it makes no sense to consider independence or identically distributed assumptions.

(2) Without additional information (more replicated chronologies or simulations) we have no reason to expect that any of the simulated quantities are approximately normally distributed. For example, we know that the Set A numbers are approximately uniformly distributed. (3) As mentioned previously, the times in Set C are influenced by the same causes and so have built-in dependence. For example, P1 takes a longer than average time to vote creating a delay for the voters that follow. Since all the voters that follow are influenced by this event, their waiting times likely correlate.

(4) Defensible confidence intervals are based on approximately IID normally distributed estimates. With only six replicates we probably do not have sufficient data for a reliable normality testing, e.g., using the KS test from Chap. 3. As a result, it is advisable to create batches. For example, we could construct three batch averages: $(0.5)(8.5 + 22.1) = 15.3$, $(0.5)(7.1 + 3.7) = 5.4$, $(0.5)(8.2 + 9.2) = 8.7$ (times in minutes). These three averages result in the 95% interval: $(-2.72, 22.32)$ minutes. Clearly, the negative time in the interval is itself an indication that the number of replicates is not sufficiently large such that a defensible confidence interval on the mean or expected waiting time can be constructed. We generally base simulation estimates on 20 or more batch averages.

4.8 Problems

1. What does IID stand for?
2. What is a Poisson process?
3. What is the CLT?
4. What is nonstationary process?
5. Assume that X is exponential with mean value 10 years. Why might this distribution be a good fit for the life of electronic components? Explain.
6. Provide an example of a service process that is exponentially distributed to a good approximation.
7. Assuming that X is exponential with $\lambda = 0.2$ per min and 0.3, 0.6, 0.1, and 0.8 are pseudorandom $U[0, 1]$, use these numbers and Monte Carlo to estimate the mean of X. Compare your estimate with the true mean.

8. Assuming that X is exponential with $\lambda = 1.5$ per min and 0.8, 0.2, 0.3, and 0.7 are pseudorandom $U[0, 1]$, use these numbers and Monte Carlo to estimate the mean of X. Compare your estimate with the true mean.

9. A simulation replicate generates a series of waiting times given by 0.0, 0.0, 5.3, 10.2, 12.2, 9.0, 8.3, 5.2, 3.2, 0.5, and 0.0 (input). What are the issues associated with using the: (1) sample mean and (2) sample standard deviation of these numbers to estimate the expected waiting time?

10. Consider the following LCG:

$$Z_i = (aZ_{i-1} + c) \mod m, \quad U_i = Z_i/m$$

with $a = 25$, $m = 1024$ and $c = 7$ and $Z_0 =$ the number corresponding to the first letter of your first name in the alphabet (e.g., Cary Grant would have $Z_0 = 3$). Also, consider the simple system:

EXPO(5 + (Last 4 Digits/1000))

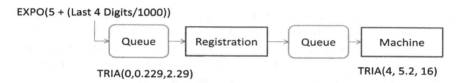

TRIA(0,0.229,2.29)   TRIA(4, 5.2, 16)

where "Last 4 digits" are the last four digits of your SSN number.

(a) Use the above numbers and Excel to simulate two replicates of a 30 min election day (polls close 30 min after opening and no new voters can enter thereafter). Make sure your answer includes figures giving screen shots of your Excel sheets with a body of text, captions, and references to your figures in the text.

(b) Give a confidence interval for the average total waiting time in queue.

(c) Suppose we are considering using the average of the first three simulated voters in place of the averages used in part b. Does the central limit theory guarantee that our sample mean is an unbiased estimate of the true expected total wait for voters (in queue)? Why or why not?

11. What is the relationship between the batch size and the normal distribution and why?

12. Use a professional software package to simulate the voting system in question 10 but this time with a 13 h election day. Run 200 replicates and put the resulting 200 total waiting times into an Excel spreadsheet.

(a) Perform input type analysis on these 200 numbers. Which distribution from the Input Analyzer minimizes the sum of squares error based on the "Fit All" algorithm?

(b) Batch or group the 200 waiting times into 20 groups of 10 and generate the sample averages. Perform input type analysis on these 20 averages (of average total waiting times). What distribution minimizes the sum of squares error based on the "Fit All" algorithm?

(c) Create a confidence interval based on the 20 averages from part b.

(d) Comment on the defensibility of the interval in part c.

13. Assume the following data are batch means each representing $n = 2$ replications. Are the results ready to present the decision-maker with a confidence interval? Recommendations?

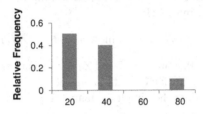

14. Consider a Poisson process with $\lambda = 0.5$ per min and exponential service times with mean 2.0 min. Assume that the system is open for arrivals only during a 20-min window. Use a spreadsheet and manual controller to simulate two replicates. Start with a high quality PRN stream of $U[0, 1]$ deviates.

15. Compare push and pull systems from a finished goods and throughput point-of-view, i.e., number of units produced in a given time period. Assume that both have the same service distributions.

16. Identify a hypothetical real-world simulation application. Assess each of the listed verification and validation techniques listed in the chapter with respect to their subjectively assessed cost and value in relation to your case.

Chapter 5
Output Analysis

After input analysis, model building, and model validation, decision support is not immediately available. The simulation team simply has a model to predict outputs or responses for given combinations of input or factor settings. Showing related animations and the results from a single system is rarely sufficient. While the process of building the model itself likely lead to insights and valuable data, using *creativity* to generate alternative systems to be evaluated is almost always *critical* to the success of the project. By single "system" we mean one combination of numbers of machines, staffers, staff schedules, and other factor levels which could, e.g., represent the current operating conditions.

The inspiration for selecting alternative systems for evaluation can come from the Theory of Constraints and lean production which are described in Chap. 7. The following is a list of example types of system alternatives that were identified and evaluated using simulation leading to positive outcomes:

- In a manufacturing project, simulation identified that one cell was both the bottleneck (slowest process that determined the system capacity) and that it was periodically starved. This was achieved by studying alternative systems having higher service rates for different machine cells. Design of experiments (DOE) methods described in this chapter helped demonstrate that only one cells reduced service rate strongly affected overall throughput and constrained the overall system. This insight led to the inspiration of combining another cell with this cell. Using the added resources, the simulation showed, the bottleneck could be sped up for all operations and would almost never be starved.
- In a health carehospital project, alternative ways to schedule nurses and patients led to alternative systems. The simulations of these alternatives showed that some choices, such as decreasing the buffer times between patients and further staggering of the arrival of nurses, would likely lead to patients achieving times closer to their target times and an ability of the unit to handle larger volumes without adding nurses.

T. T. Allen, *Introduction to Discrete Event Simulation and Agent-based Modeling*,
DOI: 10.1007/978-0-85729-139-4_5, © Springer-Verlag London Limited 2011

- In an election project, simulation predicted that, because of the slower service times of Help America Vote Act compliant machines, additional machines or lower turnout on Election Day was needed or long lines would result in 2008. This warning helped county officials to dramatically increase absentee voting through advertising spending and newspaper articles, avoiding the long lines predicted by the simulation. Also, the simulation showed that allocating additional machines to locations with longer ballots would greatly reduce the inequality in the waiting times across the county. The improved "utilization-based" formula was implemented and increased equity. This case study is described in detail in Chap. 7.
- The US military used simulation to determine the number of troops, airplanes, and ships needed to thoroughly "overmatch" Panamanian president Manuel Noriega in his extradition. By trying various combinations, military offers could see the minimum number of resources needed to quickly pacify their opponent. Thus, bloodshed was minimized and objectives were achieved.

In at statistical and technical sense, output analysis involves using the validated simulation model to prepare information to support decision-making. It involves testing alternative ways to operate systems and ensuring that the comparison conclusions are statistically defensible. Often, alternatives derive from management dogmas such as lean engineering or theory of constraints as described in Chap. 7. In some cases, large numbers of alternate approaches are generated and analyzed using formal optimization as described in this chapter.

In general, teams that are charged to consider larger numbers of factor combinations or "alternative systems" have a higher chance of discovering valuable information. In operations research terminology, this follows because their search space is less constrained and therefore more likely to contain a desirable solution. Yet, searching widely using discrete event simulation involves addressing at least two major challenges:

1. Simulation models estimate response values with Monte Carlo or replication errors as noted in Chap. 4. In other words, the codes are "noisy" system evaluators as compared with "deterministic" evaluation in standard linear and nonlinear optimization problems. This noise means that simulators need to guard against one system having a "lucky" set of replicates causing the discounting or discarding of superior solutions, i.e., system alternatives having truly higher mean performance.
2. Discrete event simulation codes can be "slow" even with modern computer power. For example, one replicate of the monthly plant operations could require 30 min at a major automotive manufacturer. Therefore, a reasonable approach applying 20 replicates would require 10 h or more for evaluating a single combination of factor settings. Compare that with 0.0001 s required by many linear programs for evaluation of the objective function.

This chapter describes several approaches to develop reliable decision support information addressing the noisy and slow challenges associated with discrete event simulation codes. In Sect. 5.1, "multiple comparison techniques" are presented which permit users to evaluate a moderate number of systems simultaneously, limiting the chance that a "lucky" system will distort quality judgments. Section 5.2 describes two types of selection and ranking methods designed to efficiently limit the confusion from the noisy simulations.

Section 5.3 describes approaches for effectively comparing tens or hundreds of systems with a small number of replications. These comparisons are based on so-called "DOE" "metamodeling" approaches, where the constructed metamodels are relatively inaccurate but computationally efficient surrogates or "stand-ins" for the discrete event simulation code. Plotting and optimizing the meta-models often provides the information that supports the final design decisions.

Section 5.4 reviews simulation optimization techniques from the recent research literature. These methods are designed to automatically search through huge numbers of alternative, systems while applying desirable numbers of replicates in each evaluation. Each optimization technique is associated with advantages for certain types of problems and addresses the challenges of noisy and slow function evaluations differently. Included with the review is a description of the proposed "population indifference zone" (PIZ) search method. Approaches like PIZ divide the huge search space into relatively small sets of solutions that efficient methods from sequential analysis can address.

In Sect. 5.5, the chapter closes with a brief description of the challenges of "steady state" simulations designed to model systems over the long run. In such systems, each replicate generally requires significant CPU time. Many methods have been developed to extract multiple outputs from these long replicates.

5.1 Multiple Comparisons Techniques

In Chap. 2, the approach for constructing confidence intervals including "half widths" was described. In Chap. 4, the batching for normality procedure was presented largely to satisfy the conditions needed for obtaining accurate confidence intervals. Applying both procedures together results in our ability to accurately predict the probability that the true mean or expected value is within one half width of our Monte Carlo estimate or simulated average. Next we need statistical results pertinent to simultaneous conclusions about two or more systems.

5.1.1 The Bonferroni Inequality and Simultaneous Intervals

Let us digress briefly for a review of probability theory. Consider two events A and B and their intersection $A \cap B$. By intersection we mean the event in which both

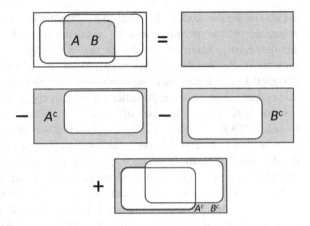

Fig. 5.1 Proof of the relationship of probabilities and the Bonferroni inequality

A and B occur simultaneously. We also consider the complement events A^c and B^c. For example, A^c is the event that A does not happen. A general result is:

$$\Pr(A \cap B) = 1 - \Pr(A^c) - \Pr(B^c) + \Pr(A^c \cap B^c). \qquad (5.1)$$

The proof for (5.1) is given by the Venn diagram in Fig. 5.1. Areas in Venn diagrams correspond to probabilities. The probability associated with the entire area is 1.0. Each box corresponds to a term in Eq. 5.1. Also, the final term corresponds to a correction for the double counting of the intersection of the complement events that would otherwise occur. From Eq. 5.1 the Bonferroni inequality follows immediately as:

$$\Pr(A \cap B) \geq 1 - \Pr(A^c) - \Pr(B^c) \qquad (5.2)$$

since $\Pr(A^c \cap B^c)$ is either 0.0 or a positive number. The key benefit of the Bonferroni inequality in Eq. 5.2 is that it holds regardless of whether the events A and B are independent or other assumptions about these events.

Next, let A be the event that the true mean or expected value for system 1 is in its $100(1 - \alpha_1)$th percentile confidence interval derived by simulation. Similarly, let B be the event that the true mean or expected value for system 2 is in its $100(1 - \alpha_2)$th its confidence interval. The probability that both means are in their confidence intervals is therefore bounded by the Bonferroni inequality.

Specifically, the probability that both means are in their confidence intervals is bounded by:

$$\Pr(\text{both means are simultaneously in their intervals}) \geq 1 - \alpha_1 - \alpha_2. \qquad (5.3)$$

To gain additional intuition about this, it might be helpful to inspect Fig. 5.2. An intuitive way to think about this relates to the worst way in which events could conspire against our complete success. We would like to have both means inside their intervals. Any other possibility is negative for us. The worst possibility occurs if one is never outside its interval while the other is. In general, having both outside their intervals might seem pretty unlikely. If we ignore its possibility, then the inequality in Eq. 5.3 changes to equality.

Fig. 5.2 Depiction of the
events and their probabilities
relating to two systems

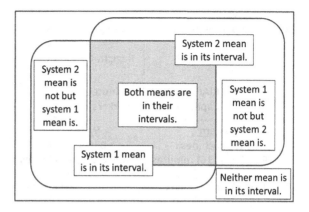

Next, let us generalize to q systems. Let μ_i refer to the true mean for the ith system and I_i refer to the $(100)(1 - \alpha_i)$th percentile interval for that system. The Bonferroni inequality gives us:

$$\Pr\{(\mu_1 \subset I_1) \cap (\mu_2 \subset I_2) \cap \ldots \cap (\mu_q \subset I_q)\} \geq 1 - \sum_{i=1,\ldots,q} \alpha_i. \tag{5.4}$$

Therefore, simultaneous confidence intervals can be derived if we set all individual α values equal to:

$$\alpha_i = (\alpha)/(q) \quad \text{for } i = 1, \ldots, q. \tag{5.5}$$

For example, assume $q = 10$ system alternatives being compared and it is desired to have a probability greater than 95% that all are in their interval. Therefore, $\alpha = 0.05$. If we use $\alpha_i = (0.05)/(10) = 0.005$ for each interval, Eq. 5.5 guarantees that the simultaneous probability of 95% is achieved. Note that the Bonferroni formula is often very conservative. Alternative simultaneous intervals procedures include those derived from Scheffe's methods and Tukey's methods.

5.1.2 Which Election System Reduces Voter Waits?

As an example of simultaneous intervals, consider the hypothetical election system shown in Fig. 5.3. The example is hypothetical since the author has not as yet been asked to evaluate election machine purchasing decisions. Therefore, the time distributions are hypothetical but potentially suggestive of actual decisions faced by election officials. The scope of the project focuses on a typical large precinct simulated for a 13-h election day. The alternatives represent different types of machines having different cost-versus-service rate trade-offs. The intent is that all options would cost the same.

For example, system 1 might represent somewhat inexpensive machines with a personal computer type small screen. The second system (system 2) might

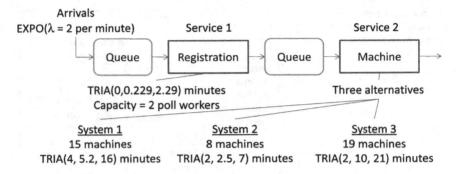

Fig. 5.3 Work flows of three hypothetical systems to be compared using simulation

represent the application of relatively costly "full-face" machines that permit the voter to see all that is being voted on at a single time. The third hypothetical system (system 3) also would correspond to personal computer type direct recording equipment (DRE) machines.

For simplicity, the following limitations were imposed in the discrete event simulation models used to evaluate the three simulations in this example:

- Time limits were applied crudely in that arrivals were assumed to begin just as the polls opened at 6:30 am and at the 7:30 pm closing time all voting terminated. In real systems, voters usually line up before the opening and everyone in the system at closing time is permitted to vote.
- Arrivals were assumed to be homogeneous. In a real problem, a non-homogeneous arrival process would be applied likely based on an empirical distribution to address rush periods.
- Only nine replicates were generated so that all numbers fit neatly in small tables. In an actual analysis, at least 20 replicates would be applied initially.

The outputs from the three simulation models are shown in Table 5.1. The table shows the average waiting times of voters summing times in registration and before voting using DRE machines. The results for the nine replicates are batched so that the resulting quantities are relatively normally distributed. Then, the half widths and simultaneous Bonferroni half widths are calculated. The calculations use the standard formula in Eq. 2.7 using $\alpha = 0.05$ and $\alpha = 0.017$ in the formulas, respectively.

The individual and simultaneous confidence intervals are shown in Fig. 5.4. From the individual confidence intervals system 2 offers significantly better performance than either system 1 or system 3, which cannot be distinguished. Yet, with the more appropriate simultaneous intervals, the only significant difference observed is that system 2 offers significantly reduced average waiting times compared with system 3. Other differences are not significant. In a real case, we might perform 100 replicates of each system and 20 batch averages. Likely then we could rank the three systems fully with statistical backing.

Table 5.1 Average waiting time outputs (minutes) describing the hypothetical election systems

Replicate or batch	System 1	System 2	System 3
Rep. #1, batch #1	39.7	3.1	46.1
Rep. #2, batch #1	36.6	3.0	33.3
Rep. #3, batch #1	50.9	4.4	44.2
Rep. #4, batch #2	67.0	2.6	49.8
Rep. #5, batch #2	52.0	4.2	37.1
Rep. #6, batch #2	46.4	5.2	69.8
Rep. #7, batch #3	37.4	4.9	61.9
Rep. #8, batch #3	35.7	2.3	50.9
Rep. #9, batch #3	19.3	6.0	47.6
Batch of size three average #1	42.4	3.5	41.2
Batch of size three average #2	55.1	4.0	52.3
Batch of size three average #3	30.8	4.4	53.5
Monte Carlo estimate (Xbar)	42.8	4.0	49.0
Batch standard deviation	12.2	0.4	6.8
Half width (95%)	30.2	1.0	16.8
Simultaneous half width (95%)	53.8	1.9	29.9

Fig. 5.4 a Individual and
b simultaneous intervals for
election example

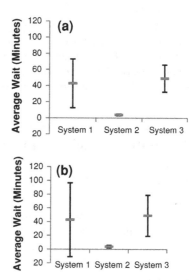

5.1.3 Sample Size Estimates

The sample size needed depends on our objectives. In the preceding example, three batches of size three constituted an acceptable sample size if our only objective was to determine whether system 2 or system 3 was better using simultaneous intervals. Yet, sample size formulas are often expressed in terms of a target or desired half width, h_0. Related formulas generally depend on the estimated standard deviations, s, derived from n_0 initial samples or evaluation

Fig. 5.5 Example spreadsheet for sample size calculation via iteration

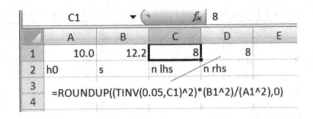

evaluations. The number of samples could represent either the number of replicates or the number of batches depending upon which the standard deviation, s, is based. Solving Eq. 2.7 for n gives us:

$$n = \text{roundup}\left[\left(t_{\alpha/2,n-1}\right)^2 (s)^2 \Big/ (h_0)^2\right]. \tag{5.6}$$

Consider that both sides of Eq. 5.2 depend on n. Therefore, it must be solved numerically. This can be achieved by putting the right-hand-side in a cell in Excel and having the formula depend on the left-hand-side. By iteration, it is possible to find a value that makes both cells equal and therefore satisfies the equality.

For example, assume that we are attempting to find the number of batch averages needed, n, to achieve a half width of $h_0 = 10.0$ and our sample standard deviation of batch averages is 12.2. By iteration, we derive a recommended sample size of $n = 8$ batch averages. Assume that $n_0 = 3$ were already complete. Then, we would need only $n - n_0 = 5$ additional batch averages to expect to reach our desired half width. Figure 5.5 shows the spreadsheet used for this calculation.

5.2 Statistical Selection and Ranking Methods (Optional)

Statistical selection and ranking methods involve the selection of a subset of systems being compared with desirable properties. Related methods are generally based on the assumption that samples from each system are normally distributed to a good approximation. Therefore, in the context of simulation it is generally desirable to derive samples that are batch averages from a batching for normality procedure as described in Sect. 3.4. Deriving desirable subsets using simulation is the subject of entire textbooks (e.g., Bechhofer et al. 1995).

5.2.1 Subset Selection and Indifference Zone Procedures

In general, selection and ranking procedures divide into two types: "subset selection" (SS) and "indifference zone" (IZ) methods. Subset selection procedures derive sets of solutions (subsets) of random size but offer the possibility of stronger quality guarantees. Indifference zone procedures derive subsets of preset sizes but offer relative weak guarantees. Guarantees are generally expressed in terms of a user declared indifference parameter, δ. The user must be willing to lose the best solutions (with the highest or lowest mean) as long as a solution with mean within δ of the best mean is retained.

For example, imagine that there are 100 alternatives and the first system (unknown to us) has the lowest mean or expected value equal to 0.0. Assume we are minimizing. The next best system (unknown to us) has a mean equal to 2.0. All other solutions have mean 10.0. We declare that we are indifferent to not finding the best solution as long as we retain at least one solution with mean within $\delta = 3.0$ of the best. Therefore, as long as the selection and ranking procedure terminates with a subset containing either system 1 or system 2 or both, we would be satisfied with the subset quality.

Selection and ranking procedures guarantee the attainment of subsets with satisfactory quality with probabilities greater than $P*$. The user can set $P*$ similar to setting α values in hypothesis testing. Such guarantees depend on the assumption of normally distributed samples or observations. In general, the smaller the δ we insist on, the higher the number of samples needed to achieve small subsets. Subset procedures can be applied with arbitrary sample sizes and $\delta = 0$, while guaranteeing subset quality. However, the final subset might be the original set of alternatives, i.e., no alternatives are eliminated.

Here, two methods are selected somewhat arbitrarily for presentation. Both apparently offer competitive properties even considering the wealth of methods developed subsequently. The first is a subset selection method from Goldsman et al. (1999). It is similar to the classical methods from S.S. Gupta and others. The second method is the "indifference zone" (IZ) restricted subset selection method from Sullivan and Wilson (1989). This second procedure is relatively complicated but it offers a rigorous guarantee and permits the selection of a subset.

5.2.2 Subset Selection

The following method achieves a subset having a probability great than $P*$ of containing a solutions with mean or expected value within δ of the optimal or best system solution. The indifference parameter, δ, can be set equal to zero if desired, but generally fewer systems will be eliminated. Let k denote the number of alternative systems being compared. Assume we are trying to find a subset containing a system with mean within δ of the smallest, i.e., we are minimizing.

Step 1. Evaluate all systems using n_0 samples, which are generally batch sample average values. Denote the resulting values $X_{i,j}$, with $i = 1, \ldots, k$ referring to the system and $j = 1, \ldots, n_0$ referring to the sample.

Step 2. Calculate the sample means or Monte Carlo estimates, $X_{\text{bar},i}$, for each system $i = 1, \ldots, k$. Denote the index for the system with the best mean as "b".

Step 3. Next, we examine the differences between the samples from each system paired with the samples from system b. Computer the standard deviations of the differences using:

$$S_{i,j}^2 = \left\{ \sum_{j=1,\dots,n0} \left[X_{i,j} - X_{b,j} - \left(X_{\text{bar},i} - X_{\text{bar},b} \right) \right]^2 \right\} \bigg/ (n_0 - 1) \qquad (5.7)$$

and

$$W_{i,b} = \left(t_{v,n0-1} \right) \left(S_{i,b} \right) \bigg/ \left[(n_0)^{1/2} \right] \qquad (5.8)$$

with

$$v = 1 - (P*)^{[1/(k-1)]} \qquad (5.9)$$

and where "t" is a critical value for the t-distribution described in Chap. 2 and Table 2.1.

Step 4. Form the subset by including only the systems with means satisfying:

$$\left(X_{\text{bar},i} - X_{\text{bar},b} \right) \leq \text{maximum} \left(W_{i,b} - \delta, \, 0.0 \right) \qquad (5.10)$$

where the "maximum" implies taking the larger value which could be 0.0.

Note that the above method with $\delta = 0$ is essentially creating single sample differences of each subsystem and the apparently best system. The approach essentially creates these difference and t-tests whether the differences are statistically significant.

As an example, consider the $k = 3$ system problem in Table 5.1 and the assumption $P* = 0.95$. This value 0.95 corresponds to a value $\alpha = 0.05$. System 2 has the smallest batch mean so $b = 2$. From Eq. 5.7, we derive $S_{1,2}^2 = 152.7$ and $S_{3,2}^2 = 40.7$. These are simply the sample variances of the differences between the three batch means of systems 1 and 3 and the batch means of system 2. Applying Eq. 5.9, we derive $v = 0.025$ from Eq. 5.9.

Applying Eq. 5.8, we derive $W_{1,2} = 30.5$ and $W_{3,2} = 15.7$. The mean differences $(X_{\text{bar},1} - X_{\text{bar},2}) = 38.8$ and $(X_{\text{bar},3} - X_{\text{bar},1}) = 48.3$ and both are larger than the respective numbers on the right-hand-side of Eq. 5.10, even with $\delta = 0.0$. Therefore, the final subset contains only system 2; others are eliminated. Also, the above method is more powerful than simultaneous intervals. The given initial set of options is {system 1, system 2, system 3}, the resulting subset is {system 2}. It is the best system with probability greater than 95%.

5.2.3 An Indifference Zone Method

This method is from Sullivan and Wilson (1989). Its objective is to start with a set of k alternative systems and terminate with a subset of m, where the method user picks m. The user picks m together with the indifference parameter d and the lower bound on the quality probability P^*. For example, one might start with $k = 100$ systems and plan to end with ten systems with one having a mean within $\delta = 3.0$ of the true best mean from the original 100 with probability $P^* = 0.95$. The procedure is based on pre-tabulated Rinott's constants, denoted here Rinott$_{k,m,n0,P*}$. Rinott (1978) was probably the first to tabulate these constants.

Step 1. (First stage) Evaluate all systems with n_0 samples. These are generally batch means with typical initial values equal to $n_0 = 10$ or $n_0 = 20$. Fewer samples are generally needed if the batches are large. Calculate the sample means, $X_{\text{bar},i}$, and sample standard deviations, S_i, for all system responses.

Step 2. Calculate the number of follow-up samples for each system using:

$$n_i = \text{maximum}\left\{ n_0 + 1, \text{roundup}\left[\left(\text{Rinott}^2_{k,m,n0,P*} \right) \left(S_i^2 \right) \Big/ \left(\delta^2 \right) \right] \right\}. \quad (5.11)$$

Step 3. (Second stage) Perform the additional n_i runs and then calculate the means of these second stage runs, $X^{(2)}_{\text{bar},i}$. Denote the index for the system with the best mean as "b".

Step 4. Using the first stage standard deviations, S_i, calculate:

$$W_i = \frac{n}{n_i}\left[1 + \sqrt{1 - \frac{n_i}{n}\left(1 - \frac{(n_i - n)\delta^2}{h^2 S_i^2} \right)} \right] \quad \text{for } i = 1,\ldots,k, \quad (5.12)$$

where h is the appropriate Rinott's constant (see below) and keep m subsystems with among these with the smallest $W_i X_{\text{bar},i} + (1 - W_i) X^{(2)}_{\text{bar},i}$ values.

The above procedure is essentially the same as the Koenig and Law (1985) procedure. Table 5.2 shows Rinott's constants for three scenarios. The first two are based on relatively small initial sample sizes, $n_0 = 10$. The third is based on a relatively large initial sample size, $n_0 = 50$. Often in the literature the intent is for

Table 5.2 Rinott's constants for: (a) $3 \to 1$, (b) $n_0 = 10$, $100 \to 10$, and (c) $n_0 = 10$, $100 \to 10$	(a)		(b)		(c)	
	k	3	k	100	K	100
	m	1	m	10	M	10
	n_0	10	n_0	10	n_0	50
	$P*$	H	$P*$	h	$P*$	H
	0.95	3.1	0.95	3.3	0.95	3.1
	0.96	3.2	0.96	3.5	0.96	3.2
	0.97	3.5	0.97	3.7	0.97	3.3
	0.98	3.8	0.98	3.9	0.98	3.5
	0.99	4.1	0.99	4.3	0.99	3.8
	0.995	4.9	0.995	4.9	0.995	4.1
	0.9995	5.6	0.9995	5.5	0.9995	5.0

indifference zone methods to provide a single solution, i.e., a subset with $m = 1$. These methods might be applied after a subset selection method has narrowed the field by eliminating low quality options. Table 5.2 shows constants relevant to starting with three alternative systems and sampling to pick a single solution, hopefully with its mean within δ of the true best mean.

The values in Table 5.2 were derived by simulation of the indifference zone procedure in the least favorable configuration using 5,000 replicates. The least favorable configuration has one mean δ higher than all others which have equal means. A binary search determined the Rinott's value that yielded the desired probability, $P*$.

5.3 Design of Experiments and Main Effects Plots (Optional)

Discrete event simulation models can be treated like any other experimental system. All of the methods from the study of experimental design or "DOE" can be used to help decision-makers understand how changing factors affects average or expected response values. These methods are described in such textbooks as Montgomery (2008) and Allen (2010). DOE methods create approximate stand-ins or surrogate "meta-models" for the relatively complicated and slow simulation models.

As an example of a widely used DOE method, consider the application of the 8 run fractional factorial experimental design in Table 5.3. The array is orthogonal in part because the columns correspond to orthogonal vectors, i.e., multiply two columns and sum and the result is zero. For example, multiplying columns A and B we derived $(-1)(-1) + (-1)(-1) + (-1)(1) + \cdots + (1)(1) = 0$. This array is also called the L8 orthogonal array by G. Taguchi and it has many desirable properties.

To apply the array in Table 5.3, we identify seven or fewer factors whose effects we would like to study. Table 5.4a shows the factors and levels for a purely hypothetical study of an election system. In the hypothetical study, the analyst has only six factors of current interest. Table 5.4a shows the array in engineering units. The -1s in the array in Table 5.3 have been replaced by the low level of the corresponding factor. The +1s have been replaced by the high level. Each row or "run" therefore corresponds to an alternative system.

	A	B	C	D	E	F	G
Table 5.3 An 8 run regular fractional factorial array also called the L8 orthogonal array	-1	-1	-1	-1	-1	-1	-1
	-1	-1	-1	1	1	1	1
	-1	1	1	-1	-1	1	1
	-1	1	1	1	1	-1	-1
	1	-1	1	-1	1	-1	1
	1	-1	1	1	-1	1	-1
	1	1	-1	-1	1	1	-1
	1	1	-1	1	-1	-1	1

Table 5.4 (a) The factors and levels and (b) the scaled experimental design in engineering units

(a)

Factor	Low	High
A. #Reg. workers	2	3
B. #Voting machines	15	17
C. Parm. b voting dist.	16	18
D. Parm. m voting dist.	5.2	7.2
E. #Registration queues	1	2
F. Arrival rate (#/minute)	2	2.3

(b)

System	A	B	C	D	E	F	G	Avg. Wait
1	2	15	16	5.2	1	2	NA	8.3
2	2	15	16	7.2	2	2.3	NA	10.4
3	2	17	18	5.2	1	2.3	NA	36.9
4	2	17	18	7.2	2	2	NA	28.4
5	3	15	18	5.2	2	2	NA	17.2
6	3	15	18	7.2	1	2.3	NA	18
7	3	17	16	5.2	2	2.3	NA	20.7
8	3	17	16	7.2	1	2	NA	22.6

Fig. 5.6 Main effects plot for hypothetical election systems example

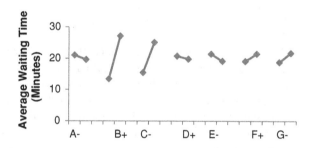

For example, the first system has two registered workers, 15 voting machines, the service times are TRIA(4,5.2,16), there is a single queue before registration, and the arrival rate in the precinct is assumed to be 2.0 voters per minute. Usually, one randomly reorders the runs before performing the experiments but this is typically not needed in discrete event simulation studies. In a real example, eight simulation models would be created and run. The response is typically the average of 20 or more replicates but large numbers of replicates are generally not needed because there is a pooling effect. Hypothetical responses for the batch averages are shown on the right-hand-side of Table 5.4b.

Once the responses have been collected, there are many options for analysis including hypothesis testing. Here, we focus only on one widely used type of plotting called "main effects plotting". Figure 5.6 shows the main effects plot for the data. Each point represents the average of four responses with the relevant factor either at the low or high level. For example, the first point in the plot is the (8.3 + 10.4 + 36.9 + 28.4)/4. Main effects plots are reasonably easy to create using standard spreadsheet software. Figure 5.6 was created using the Microsoft® Excel "Line" chart feature from the "Insert" menu. The term "effect" here refers to the average of responses at the high side minus the average at the low side for each factor.

In general, almost every factor has an effect over some range. The plot tells us which factors have major effects over the range studied. In this hypothetical case, the response is the average waiting time. If the data were real, we could see that the average waiting time is highly sensitive to changes in the numbers of machines and the parameter b from the triangular distribution. Other factors have relatively small (if any) effect over the levels studied. This information could aid officials so that they focus on the most appropriate issues. In this case, they would focus on additional machines and driving down the voting times, e.g., through voter preparation.

Main effects plots provide a visual indication about the effects of factors on system responses. Such information can help because factors can represent decision variables or, alternatively, assumption parameters. Yet, main effects plots do not, by themselves, constitute statistical proof and do not even visually account for multiplicity issues. Probably the most common approach for hypothesis testing is to use a normal probability plot of the estimated effects. Estimate effects are the differences between the averages at the high side minus the averages at the low side for each factor. For example, in Fig. 5.6 the largest effect is factor B which is +13.7.

The idea behind normal probability plotting is simple. If the data is all noise with no pattern, then all factors would have no effect. Also, then the effect estimates will be approximately normally distributed as can be predicted using the central limit theorem in Chap. 2. IID normally distributed random numbers line up on normal probability plots. Therefore, significant effects will be large in magnitude and not on the line established by the effects associated with inert factors. Figure 5.7 shows the normal probability plot for the data in Table 5.4b. The effects from factor B (#Voting machines) and factor C (Parameter b voting distribution) are not only apparently large in the main effects plot. Also, these effects are proven significant by the normal probability plot of effects. Admittedly, the line in the plot is drawn with some subjectivity.

Standard software such as Minitab and JMP generate normal probability plots of estimated effects automatically. Figure 5.7 shows the steps in excel for generating the $(i - 0.5)/$(number of effects) column, the sorted effects, and the Φ^{-1} column. Here, Φ^{-1} is the cumulative inverse (F^{-1}) for the normal distribution. This approach can be applied for experiments design using the standard $n = 8$ run, $n = 16$ run, and $n = 32$ run regular fractional factorials. Also, the standard method uses the estimated effects for all $(n - 1)$ columns even if fewer than $(n - 1)$ factors are used. The reason is that using as many numbers as possible that are likely normally distributed helps to establish the line more accurately.

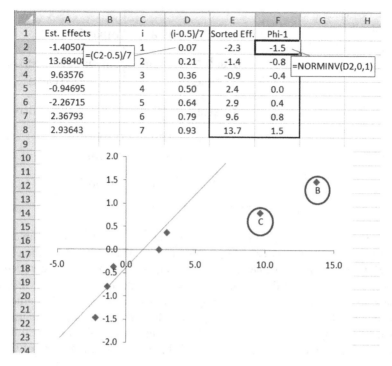

Fig. 5.7 Normal probability plot of estimated effects

5.4 Black Box Simulation Optimization Methods (Optional)

The literature on simulation optimization is substantial and growing. See Fu et al. (2005) for a recent literature review. Here, we divide methods by two issues. First, some methods are specific for certain types of simulation and others are "black box" or generic. Generally, specific methods that exploit the problem structure identify desirable system design alternatives in far less computation time. Yet, black box methods are applicable to virtually all types of problems and require limited expertise for their application. Second, some methods use constant sample size and others use variable sample sizes.

One way to characterize the types of simulation optimization methods is shown in Fig. 5.8. Simulation optimization is of increasing importance in part because, with modern computers, more accurate and influential simulation models are being built and used than ever before. Some of these models are part of automatic control systems such that they generate predictions leading, through optimization to immediate actions. The intent in drawing Fig. 5.8 is to clarify the author's subjective view about the history, the current state of practice, and research trends in the field of simulation optimization.

It is perhaps surprising that commercial software vendors have been slow to use variable sample size methods. A well-known commercial product called

Fig. 5.8 Concept map to
characterize the types of
simulation optimization
methods

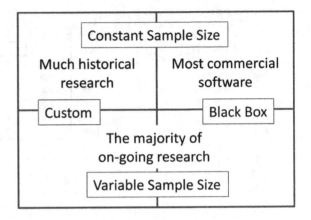

"OptQuest" uses them in a limited way. Solutions are evaluated with sample sizes sufficient to show that they are comparable or worse than δ way from the current best solution in expected value of response. Yet, how sample sizes are varied for comparisons with other design alternatives is a little unclear. In the research literature, however, the benefits of using small numbers of samples to eliminate poor solutions have been clear for many years. It is likely true that some users of commercial software are waiting many times longer than needed and still derive lower quality solutions than what could be obtained using state-of-the-art techniques.

In addition, a promising set of black box techniques attempts to address the inherent limitations of all types of simulation. By considering the possibility of developing models with varying degrees of detail, these "multi-fidelity" optimization techniques simultaneously work toward desirable real world factor settings while mapping out the systematic errors of the various simulation codes from which they draw evaluations. Huang et al. (2006) contains a promising multi-fidelity methods that blurs the line between DOE and simulation optimization methods.

5.4.1 Variable Sample Size Methods

In this section, a simple black box, variable sample size method is described. This method, by itself, is likely not competitive with commercial or research methods. Yet, it illustrates the combination of a population based optimization search with statistical selection and ranking methods. Combining these methods is one of the important, on-going threads in the research literature. The promise of these methods is that, by dividing the large optimization problem into smaller problems, efficient variable sample size methods from selection and ranking can guarantee that desirable solutions are not lost. These methods are promising also because they do not ask simulation to compare too many similar solutions which are difficult to tell apart with noisy and slow evaluations.

5.4.2 Population Indifference Zone Search Method

Step 1. (Initialization) Create 100 system alternatives by sampling uniformly from the decision space. For example, system 1 in Table 5.4 might be one of the random selections.

Step 2. (Subset selection) Perform subset selection with $P^* = 0.99$ based on $n_0 = 10$ batch means as described in Sect. 5.2.2.

Step 3. If the number of solutions in the subset is less than or equal to 10, go to Step 5.

Step 4. (Indifference zone selection) For the not-eliminated solutions from Step 2, perform indifference zone selection with the first stage given by the $n_0 = 10$ batch means from Step 3. Use $m = 10$, $P^* = 0.99$, and $\delta = 1.0$ where 1.0 represents a reasonably small number in the problem specific units. Note that applying the Rinott's constants for the $k = 100$ case (Table 5.2) will be conservative since the number of solutions being compared is generally fewer than 100.

Step 5. (Termination) Are any of our solutions good enough? If yes, stop otherwise continue.

Step 6. (Form the next population) Copy the ten highest ranked solutions into the next population. If the subset selection eliminated 90 or more solutions, these have the highest sample means. Otherwise, they are the subset from the indifference zone procedure. Fill the remaining 90 solutions with uniform random selections from the decision-space.

The population indifference zone (PIZ) search method has the pleasing properties:

1. (Black box) PIZ is reasonably simple and applicable to all simulation optimization problems with the only requirement being that one can sample uniformly from the search space,
2. (Efficient) since poor quality solutions are generally eliminated from consideration in Step 2, they are not permitted to use excessive amounts of CPU run time, and
3. (Bounded) the chance that PIZ finds a good solution and loses it is bounded. After ten populations, the procedure should keep a solution within 10.0 units of the best it has searched with a probability greater than $1 - [(10)(0.01 + 0.01)]$ equals 80%. This result derives directly from successive applications of the Bonferroni inequality in Eq. 5.1.

In addition, perhaps the most pleasing property of PIZ is that it can form a springboard for more efficient search methods. For example, some or all of the 90 solutions in each population can be generated using a method other than uniform random selection. Heuristic approaches such as genetic algorithm (GA) mating, scatter search, and taboo search can be used to populate the new generation. Also, in some cases, few (if any) solutions will be eliminated by the subset selection.

Then, additional phases of subset selection can be added with escalating numbers of evaluations to enhance efficiency.

Yet, many black box methods are not suitable for problems involving numbers of factors greater than 100. For such case, e.g., pilot scheduling or rostering, it can be critical to exploit the problem structure. Otherwise, computers could run for days and generate no acceptable solution. For example, linear constraints associated with network flows can permit the operations researcher to address a problem effectively with many fewer decision variables. As an example of recent simulation optimization exploiting problem structure and using variable numbers of samples see Zhao and Sen (2006).

5.5 Output Analysis and Steady State Simulation

This chapter has described simultaneous intervals, selection and ranking, experimental design, and simulation optimization. Each of these relates to the manipulation of simulations and outputs to generate information to support solid judgments by decision-makers. Yet, the focus here has been on batch means derived by replications of finite duration simulations. Another important set of topics relates to modeling systems in steady state, i.e., systems with no finite duration.

In steady state simulations, practitioners often perform a single extremely long replication. Because of the high computational time needed for this long replication, it is generally desirable to extract more than a single number. Yet, as described in Chap. 4, it is difficult to generate independent, identically distributed random numbers from a single simulation run. Therefore, reliable inferences based on the central limit theorem and confidence intervals are difficult to obtain. A recent review of related literature is given by Alexopoulos (2006). In this literature, approaches have been invented to achieve approximate independence of the multiple outputs from the same replication.

In the next chapter, we focus on paper and pencil alternatives to discrete event simulation associated with queuing theory. In these cases, steady state modeling is paradoxically relatively easy compared with modeling finite horizon time periods.

5.6 Problems

1. What makes simulation optimization more challenging than ordinary "deterministic" optimization such as standard linear and nonlinear programming?
2. Will subset selection procedures ever establish significant differences that simultaneous confidence intervals fail to establish? Explain briefly.
3. With multiple systems being considered, are simultaneous intervals wider or narrower than individual confidence intervals? Explain briefly.

4. There are alternative methods for establishing simultaneous confidence intervals other than those based on the Bonferroni inequality. Name at least one way in which each of these alternatives might be more desirable for at least one situation.

5. Assume that there are three alternative systems of interest to decision-makers. The simulation batch averages and batch standard deviations based on 10 batch means for each system are $\mu_1 = 10$, $s_1 = 2.6$, $\mu_2 = 25.3$, $s_2 = 4.1$, $\mu_3 = 12.2$, and $s_3 = 3.6$. Present the results using appropriately constructed intervals.

6. Assume that there are four alternative systems of interest to decision-makers. The simulation batch averages and batch standard deviations based on 5 batch means for each system are $\mu_1 = 20$, $s_1 = 5.6$, $\mu_2 = 25.3$, $s_2 = 4.1$, $\mu_3 = 12.2$, $s_3 = 3.6$, $\mu_3 = 20.2$, and $s_3 = 5.6$. Present the results using appropriately constructed intervals.

7. Suppose you are comparing ten systems. What is the setting for each confidence interval α such that the chance that all means are in their intervals is greater than 95%? If you apply an inequality other than the Bonferroni inequality in your answer, please clarify.

8. Suppose you are comparing five systems. What is the setting for each confidence interval α such that the chance that all means are in their intervals is greater than 95%? If you apply an inequality other than the Bonferroni inequality in your answer, please clarify.

9. What is an indifference parameter, δ?

10. In this problem, we would like to find the system with the lowest average waiting time and do not want to eliminate from considerations solutions which might be the best ($\delta = 0.0$). Assume that there are four systems of interest and batch averages are as follows. All responses are in minutes. For system 1, the averages are 22, 34, 15, 7, and 23. For system 2, the averages are 51, 25, 43, 45, and 43. For system 3, the averages are 45, 30, 35, 38, and 40. And for system 4, the batch averages are 43, 52, 55, 29, and 54. Which systems (if any) can be eliminated with probability of retaining the best higher than 95%?

11. Why cannot we use $\delta = 0.0$ in the indifference zone procedure?

12. Assume that responses for three systems are generated as follows and we are trying to find a single system with true mean within $\delta = 5.0$ units of the best system with probability greater than $P^* = 0.95$. For system 1, the batch means are normally distributed with mean 30 and standard deviation 10. For systems 2 and 3, the batch means are normally distributed with mean 40 and standard deviation 20. Apply the indifference zone procedure using $n_0 = 10$.

13. The PIZ method is supposed to use variable sample sizes. Where do the variable sample sizes enter into the method?

14. Perform at least three iterations of the PIZ method to attempt to solve the "a_4" problem given by:
   Minimize: $\sum_{i=1,\dots,20} (i)(x_i^4) + \varepsilon$ with $\varepsilon \sim N[0.0, 5.0]$ {by changing $x_1, \dots, x_{20}$}
   Subject to: $-1.28 \le x_i \le 1.28$ for $i = 1, \dots, 20$. What makes this a simulation optimization problem?

15. What aspect of output analysis does Alexopoulos (2006) focus on?

Chapter 6
Theory of Queues

Queuing theory can be viewed as a by-hand alternative to discrete event simulation. Yet, for many reasons discrete event simulations are often developed simultaneously with queuing models for the same or related systems. Queuing theory models offer:

1. *Validation* for the results from complicated simulations to verify that they are in the right ball-park,
2. *Efficiency* for deriving the numbers of machines when simulations are too slow (e.g., if we have a need to develop recommendations at 800+ voting locations inexpensively), and
3. *Insight* into how decisions affect outputs that complicated simulation models cannot provide.

Queuing theory models provide these benefits with at least two types of associated costs. First, users need to make a limiting set of assumptions about arrivals and service distributions. These assumptions might not apply to any reasonable approximation in some cases of interest. Making them could lead to inadvisable recommendations. Second, queuing theory models are associated with complexity and abstract concepts. These take time to understand and to apply confidently.

This chapter is intended to review only the most widely used queuing theoretic formulas and concepts. Wolff (1989) is a textbook containing many other useful queuing formulas. Research related to queuing theory continues with topics ranging from exploiting queuing structure in optimization methods to developing efficient approximations for a wider variety of assumptions.

Before diving into the structure of queuing systems in steady state, this chapter begins with two of the most practical applications of queuing theory. In Sect. 6.1, the implications of queuing theory for the case in which all servers are 100% utilized are reviewed. Section 6.2 presents a reasonably simple formula to estimate the number of machines needed based on the work of Kolesar and Green (1998). These results are based on the limiting assumption of Poisson processes for arrivals and exponentially distributed service times. Then, Sect. 6.3 covers the

T. T. Allen, *Introduction to Discrete Event Simulation and Agent-based Modeling,*
DOI: 10.1007/978-0-85729-139-4_6, © Springer-Verlag London Limited 2011

fundamental theory of so-called "Markovian" queues in which only the present system state is relevant to predict future performance. Section 6.4 includes a recounting of "Little's Law" which relates to flow conservation. This permits the derivation of average waiting times for so-called "$M/M/c$" queuing systems.

6.1 Steady State Utilization

Steady state queuing theory focuses on system properties in the long run. Over the long run, gradual accumulations are a concern because there is the potential for massive or even infinite buildups. In a well-known saying, the economist John M. Keynes (1923) observed that the, "...long run is a misleading guide to current affairs. In the long run we are all dead." His purpose was to diminish the importance of long run or steady state results in decision-making. Yet, the steady state approximation can sometimes give relevant insights. For example, a long run voting period might exceed 10,000 h in length, which could relate to an extended early voting period. Predicting performance in time periods this long or longer can be of interest to many types of organizations.

Consider the following example which might be regarded as "middle" run, i.e., not a 13-h Election Day (short run or finite horizon) but not necessarily the long run either. Assume a long 1,000 h election period with 10,000 voters arriving randomly throughout the period. Further, assume the average voting time is 10.2 min. How many voting machines would we need?

It is tempting to calculate the total amount of time on machines needed and purchase only enough machines to match that time. This leads to:

$$\#\text{machines}$$
$$= [(10,000 \text{ voters})(10.2 \text{ min/voter})]/[(100 \text{ h})(60 \text{ min/h})] \quad (6.1)$$
$$= 17.0 \text{ machines}$$

Simulation based on exponential interarrival times and exponential service times in this case predicts 36.0 ± 14.3 min average waits with an average of 60.0 ± 23.0 voters in line. These Monte Carlo predictions for the means or expected values derive from 20 replicates (with no batching). Studying the individual replications finds a pattern of concern. In every almost every case, the queue length towards the end of the period is long and growing. If the already long 100 h period were extended to handle proportionally more voters, average waiting times would likely continue to grow.

Let λ ("lambda") denote the long run average arrival rate of entities to our system in units of entities per minute. Let μ ("mu") denote the long run average service rate of each server or machine also in units of entities per minute. Therefore, intuitively the long run average interarrival and service times must be $1/(\lambda)$ and $1/(\mu)$, respectively. Let c denote the number of machines in the system. In terms of these quantities, we can define the long run average utilization, ρ ("rho") as:

$$\rho = (\lambda)/[(\mu)(c)] \tag{6.2}$$

In our above example, one has $\lambda = (10{,}000)/[(100)(60)] = 1.667$ voters per minute and $\mu = 1/(10.2) = 0.098$ voters per minute. Therefore, with $c = 17$ machines $\rho = 1.00$ or 100% utilization.

The steady state theory developed in this chapter predicts that, over an infinite period, the relevant systems with 100% utilization develop infinitely long lines on average. Theory in Wolff (1989) extends this result to virtually any system with random interarrival times and/or service times. Apparently, the slight bunching of arrival and/or collections of slow servicing time causes lines to form and to grow to infinite size. In practice, it is not clear how long one needs to wait to experience long run performance. For example, even with a 100 h Election Day and $\rho = 1.00$, the average waits are an estimated 36.0 ± 14.3 min, not infinity minutes.

6.2 Number of Machines Formula

In this section, an approximate formula for the number of machines needed is presented developed by Kolesar and Green (1998). This formula was developed to achieve the desirable property that, in steady state, only $100p$ percent of entities have any wait at all. The fraction, p, is an adjustable parameter of the method. The formula is based on the assumption exponentially distributed interarrival times (Poisson arrivals) and c servers each requiring exponentially distributed service times.

The number of machines, c, needed is:

$$c = \text{roundup}\{(\lambda)/(\mu) + z_{1-p}[(\lambda)/(\mu)]^{1/2} + 1/2\} \tag{6.3}$$

where λ is the arrival rate, μ is the service rate of individual machines, and z is the critical value for the standard normal distribution. In terms of the inverse cumulative for the standard normal, $F^{-1}()$, we have $z_{1-p} = F^{-1}(1 - p)$. Tabulated values for z_{1-p} are shown in Table 6.1a. For example, with $p = 0.3$, $z_{1-p} = 0.52$. In the example in Sect. 6.1, we have $(\lambda)/(\mu) = 17.0$. With $p = 0.3$, we have $c = 20$ machines recommended. In the long run, approximately 30% of the voters would experience a nonzero wait. Others would have no wait. Other tabulated values from the formula in Eq. 6.3 are shown in Table 6.1b.

The assumptions of Poisson arrivals and exponential interarrival are always approximate when applying to real situations. Equation 6.3 is an approximation of a more accurate equation based on results provided later in this chapter. The region of validity of the approximation in Eq. 6.3 is roughly $p \leq 0.3$. Yet, it is perhaps true that Eq. 6.3 leads to overly conservative allocations. Election officials and others might not be concerned about the long run and simply use 17 machines. Yet, in the elections example there is another consideration. The officials are simultaneously running hundreds of precincts and need to

P	z_{1-p}
(a)	
0.3	0.52
0.2	0.84
0.1	1.28
0.05	1.64

λ/μ	c
(b)	
1	3
2	4
4	7
8	12
16	21
32	39
64	73
128	141
256	274

Table 6.1 (a) Standard normal critical values and (b) outputs of Eq. **6.3** with $p = 0.2$

guarantee equal access to equipment for all. Therefore, conservatism is probably appropriate.

6.3 Steady State Theory (Optional)

In this section, formulas that describe one specific type of waiting systems are presented. The systems here are "Markovian" (*M*), named after the Russian mathematician Andrey Markov who pioneered much related research. Specifically, the Markovian systems described here are characterized by exponentially distributed interarrival times (Poisson arrivals) and exponentially distributed service times. The notation "*M/M/c*" summarizes these assumptions with the first *M* referring to Poisson arrivals. The second *M* refers to exponential service times. The "*c*" refers to the number of machines in parallel.

The defining attribute of Markovian systems is the "Markovian property" which means that the future behavior is fully predictable knowing only the present system state. It is perhaps true that all physical systems are Markovian with a sufficiently detailed description of the present state. However, because of the memoryless property of the exponential distribution described in Sect. 6.3, *M/M/ c* queues are Markovian even with only a single state variable. That variable is the number of entities in the system either waiting or being served. Because of the memoryless property, the distribution of the time to the next arrival or service does not depend on any information about the system.

In steady state, the "flow" of entities into any state equals the flow out of that state. Otherwise, there would be build-up that would grow to infinite size over time. The term "flow" refers to the transition rate of entities in one state moving to the

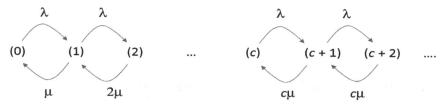

Fig. 6.1 Steady state flows into and out of each state in $M/M/c$ queuing systems

next state multiplied by p_n, the long run probability that entities will be in the state (n) being departed. Figure 6.1 shows the transition rates into and out of the states: number in system $= 0$, number in system $= 1, \ldots$ For example, in the state (1), there is one entity in the system. Arrivals are happening at a rate λ, which could drive the system to state (2). Also, service is happening at a rate μ, which could drive the system to state (0). The key point is that the flow rate downward continues to increase as more and more machines are in use until all machines are utilized.

The description here owes to Adan and Resing (2002). The situation in Fig. 6.1 leads to the flow balancing equations:

$$\lambda p_{n-1} = \min(n, c)\mu p_n \quad \text{for } n = 1, 2, \ldots \tag{6.4}$$

Rearranging and iterating results in:

$$p_n = (\{[(c)(\rho)]^n\}/(n!))(p_0) \quad \text{for } n = 0, \ldots, c \tag{6.5}$$

where $\rho = (\lambda)/[(\mu)(c)]$ is the utilization and

$$p_{c+n} = (\rho^n)(p_c) = (\rho^n)(\{[(c)(\rho)]^n\}/(c!))(p_0) \quad \text{for } n = 0, 1, 2, \ldots \tag{6.6}$$

Since all probabilities must sum to 1.0, we can calculate the probability that the long run system will be in the (0) state as:

$$p_0 = \left\{ \left[\sum_{n=0,\ldots,c-1} \{[(c)(\rho)]^n\}/(n!) \right] + (\{[(c)(\rho)]^n\}/(c!))[1/(1-\rho)] \right\}^{-1} \tag{6.7}$$

The probability from Sect. 6.2 that a job must wait, p, can now be derived exactly as:

$$p = p_c + p_{c+1} + p_{c+2} + \cdots \tag{6.8}$$

$$= (\{[(c)(\rho)]^n\}/(c!))\{(1-\rho)\left[\sum_{n=0,\ldots,c-1} \{[(c)(\rho)]^n\}/(n!) \right] + (\{[(c)(\rho)]^n\}/(c!))\}^{-1}$$

$$\tag{6.9}$$

Equation 6.9 can be solved for c to derive the exact formula for which the Kolesar and Green (1998) equation in (6.2) is a simple, easier-to-apply approximation.

As an example, assume $\lambda = 0.5$ entities per minute and $\mu = 0.1$ entities per minute. With $c = 5$ machines, the long run probability of waiting is $p = 1.0$ from

Eq. 6.9. With $c = 6$, the probability of waiting is 0.83333. This is the steady state probability that a randomly selected arrival will wait.

6.4 Little's Law and Expected Waiting Times

Little's law concerns the steady state or long term average number of entities in the system, E[number] $\equiv L$, and the long term average time these entities stay in the system, E["sojourn" time] $\equiv W$. It also involves the arrival rate, λ. This law governs all steady state queuing systems including systems with Poisson arrivals and exponential servers. The law follows directly from the assumption that the system can reach steady state, i.e., no build-ups occur because of overwhelming demands. Little's law is:

$$\text{E[number]} = (\lambda)\text{E["sojourn" time]} \rightarrow L = \lambda W. \qquad (6.10)$$

An extension of the law concerns the average waiting times, W^q, and the average number in queue, L^q:

$$L^q = \lambda W^q. \qquad (6.11)$$

For $M/M/c$ queuing systems, we calculate the expected number in line, L^q, using:

$$\text{E[number waiting]} \equiv L^q = \sum_{n=1,\dots,\infty} (n)(p_{c+n}) \qquad (6.12)$$

$$= [(p)(\rho)]/(1 - \rho) \qquad (6.13)$$

where p is given by Eq. 6.9. From Little's law and Eq. 6.11, we have:

$$\text{E["sojourn" time]} \equiv W^q = (L^q)/(\lambda) \qquad (6.14)$$

where L^q comes from Eq. 6.13.

Revisiting the same example from Sect. 6.3, assume $\lambda = 0.5$ entities per minute and $\mu = 0.1$ entities per minute. With $c = 5$ machines, the long run probability of expected number in queue (L^q) and the average number waiting (W^q) are undefined or ∞ from Eqs. 6.13 and 6.14, respectively. With $c = 6$, the expected number in queue is $L^q = 2.93$ entities. The expected waiting time is $W^q = 5.9$ min. The long run probability of a randomly selected entity waiting is $p = 0.83333$.

6.5 Summary Example

Here we consider a constantly busy fast food restaurant. The restaurant experiences rush periods but they occur at unpredictable times. Nothing particularly constrains arrivals because there is a large population of potential customers near the restaurant. These are the conditions that make the assumptions of steady state and Poisson arrivals reasonable. Even if the restaurant were only open for short periods each day, the steady state assumption might lead to interesting predictions.

Assume that customers arrive at the restaurant with exponential interarrivals with mean time of 20 s ($\lambda = 3$/min). The service time is highly random with mean time of 1 min ($\mu = 1$/min). Assume that the customer monopolizes the cashier for the whole time while the order is being processed.

The restaurant currently has $c = 4$ cashiers. What is their utilization rate and what does this utilization imply about the system? The utilization is given by $\rho = 3/[(1)(4)] = 0.75$. This rate indicates that cashiers are idle for a significant portion of their time on the job. However, this utilization might be considered reasonably high for a system characterized by so much variation. Lines likely become long and the system is not very resilient should one of the cashiers go on break or be absent.

Next, we estimate the number of cashiers needed so that the chance customers will need to wait is less than 20% ($z_{0.80} = 0.84$). How many cashiers are needed and what assumptions need to be made for your estimate? Using the Kolesar and Green approximate formula in Eq. 6.3 based on the assumptions of exponential interarrivals and exponential service times and machines in parallel, we have the following formula:

$$\#\text{cashiers} = \text{roundup}\{3/1 + 8.4[\text{sqrt}(3/1)] + 0.5\} = 5. \qquad (6.15)$$

Using (6.7) to solve for n would yield the same number. However, using the approximate formula gives us a little more intuition, i.e., we see the need for the bare minimum plus a buffer to address the variation in arrivals and service processes.

Alternatively, we could simulate with different numbers of machines and pick the system with the most desirable properties. Yet, simulation might be more time consuming, require expensive software, and not provide as much intuition. Since the steady state and Markovian (or exponential interarrivals and service distributional) assumptions are reasonable, we do not need simulation.

6.6 Problems

1. Assume that we have a detailed call center discrete event simulation model that predicts the number of expert operators needed to handle a hypothetical new type of rush period. Why might we also develop queuing models for the same purpose?
2. Assume that there are two $M/M/c$ processes in series, what are the state variables? Is the system still Markovian?
3. Assume that we have an $M/M/c$ system, we select $p = 0.2$ fraction of entities to wait, $\lambda = 0.8$ entities per minute, and $\mu = 0.1$ entities per minute. Use the Kolesar and Green approximate formula to estimate the number of machines needed.

4. Assume that we have an $M/M/c$ system, we select $p = 0.2$ fraction of entities to wait, $\lambda = 1.6$ entities per minute, and $\mu = 0.1$ entities per minute. Use the Kolesar and Green approximate formula to estimate the number of machines needed.

5. Assume that we have an $M/M/c$ system, $\lambda = 0.7$ entities per minute, and $\mu = 0.1$ entities per minute and we use eight machines. What is the exact probability that an arbitrary entity will wait? Also, what is the expected waiting time?

6. Assume that we have an $M/M/c$ system, $\lambda = 0.5$ entities per minute, and $\mu = 0.1$ entities per minute and we use six machines. What is the exact probability that an arbitrary entity will wait? Also, what is the expected waiting time?

7. Assume that we have an $M/M/c$ system, $\lambda = 0.8$ entities per minute, and $\mu = 0.1$ entities per minute and we use nine machines. What is the exact probability that an arbitrary entity will wait? Also, what is the expected waiting time?

Chapter 7
Decision Support and Voting Systems Case Study

This chapter describes practical information relevant to simulation projects. In general, the inspiration for system changes can come from many sources including from competitors. Also, the creativity involved with identifying alternative systems is generally critical to project success. Simulation is usually only useful for evaluating hypothetical changes. Without inspired alternatives, the value is limited. Dogmas like theory of constraints (Sect. 7.1) and lean production (Sect. 7.2) can provide the needed inspiration.

Also, it is generally important to consider:

1. The *financial cost of waiting time* which might relate to a multiplier times the value of the work in process or the direct cost of employing people who are waiting. Such multipliers can range from 0.02 that accounts only for the opportunity cost of investing in inventory to 2.0 which might account for loss-of-good will and lost future sales. Different organizations have different conventions for how to evaluate inventory and/or waiting costs. The implied cost of waiting can often be inferred by observing historical choices to purchase resources. For example, consider that university administrators would not have purchased 40 computers for a computer laboratory costing $100,000 including software and support costs if they placed zero value on student waiting times.

2. The *possibility of making policy changes* and not merely considering investments in additional resources. For example, putting up a sign to prepare voters showing the ballots that they will be voting on could greatly reduce the average voting (service) times. Such a subtle change would require minimal capital investment but could be effectively equivalent to having many additional machines. With signs, fewer voters are reading the entire ballots while monopolizing the existing machines. Consider also that one resource might be dedicated for entities having short processing times like a computer dedicated for printing only. These seemingly subtle policy changes can give rise to alternative systems that can be compared using simulation.

T. T. Allen, *Introduction to Discrete Event Simulation and Agent-based Modeling,*
DOI: 10.1007/978-0-85729-139-4_7, © Springer-Verlag London Limited 2011

Section 7.3 describes an actual discrete event simulation project performed on behalf of the Franklin County, Ohio Board of Elections. In Sect. 7.4, there is a supermarket case study more similar to what a student could be expected to produce in a semester long course. Exercises possibly relevant to university instruction are described in Sects. 7.5 and 7.6.

7.1 Theory of Constraints

The theory of constraints is a management dogma proposed by E. M. Goldratt and described in his book "The Goal" and subsequent books (3rd edition in Goldratt 2004). The central idea is simple. In many or all complicated systems there is exactly one bottleneck. This is a subsystem with the following property: If its capacity is increased, the overall capacity increases. All other subsystems are not bottlenecks. Increasing their capacity has little or no effect on overall system performance.

The implications for the theory of constraints for simulation projects are also clear. After a validated model has been built, the alternatives that are relevant for study relate to different ways to alleviate the flow clogging the bottleneck or to increase its capacity to handle the flow. For example, in an election system it might be tempting to focus on adding additional poll workers to support registration. However, if the bottleneck were the direct recording equipment (DRE) voting machine, the theory of constraints says that additional poll worker recruitment would be almost surely a waste of money (unless they could help speed up DRE service).

Focus on alternatives related to adding capacity to the DRE system is recommended. For example, one might simulate locations with added DRE machines or improved voter preparation through signage or hand outs. Such preparation can drive down the service times and therefore alleviate the bottlenecks. Actions that affect the bottleneck subsystem are the only ones that can increase throughput and thus key outcomes such as voter experience or, in business situations, revenues and profits. Simulation can also be used to investigate how many resources can safely be extracted from non-bottlenecks and used to alleviate bottlenecks.

7.2 Lean Production

Toyota has been among the most admired companies in the world for at least 30 years despite recent quality problems with their electronic control systems. This has occurred in part because of the many innovate aspects of the Toyota production system including those documented in Ohno and Bodek (1988) and more recently in Liker (2005). The TPS has inspired management movements including re-engineering and lean engineering. There are many aspects of the TPS. Here, we focus on five that have inspired our simulation investigations.

1. *Eliminating batch operations*: batch operations form natural bottlenecks and force producers to make quantities of similar parts regardless of what is demanded. The TPS dogma is to work toward a batch size of a single unit. For example, if you are making four sandwiches, a non-lean or "mass" producer might put eight pieces of bread in an oven, then spread peanut butter on four, then spread jelly on four, and assemble. Then, the batch size would either be four or eight. The TPS way would be to make one sandwich, then the second, the third, and the fourth. One benefit is that individual items are produced in short cycle times, i.e., they do not need to wait to be completed because their parts wait for batch processing. Therefore, customers receive their products with minimal lead times.

2. *Shortening set-up times and "mixing" production*: moving to a batch size of a single unit might seem unwise if it were not accompanied by a concerted effort to reduce the setup times of machines. For example, one advantage of spreading all the peanut butter at one time would be that the knife can be cleaned before being used on jelly. Yet, the TPS or lean producer would simply work to make cleaning extremely efficient. Then the cost of switching back and forth would be minimal. Also, the lean producer would purchase special ovens that permit one-piece-flows. Taking these actions, the lean producer would be able to switch effortlessly from producing peanut butter and jelly sandwiches to producing tuna fish sandwiches and "mix" the production. Faster set-ups and mixed production also result in generally shorter lead times.

3. *Implementing a "pull" system*: in the TPS, items are custom made precisely to match orders. Therefore, items are "pulled" into the system to meet demands and not pushed into the world to meet forecast demands or to fill out batches for batch operations. With one-piece-flows and mixed production, the item being ordered is made and delivered quickly, i.e., with greatly reduced lead time.

4. *"Leveling" demand*: the randomness of demand can hurt performance of any production system. For example, if all voters arrive at the beginning of Election Day, then waiting lines will be much longer than they need to be. Organizations can, however, take action to try to even or "level" out their demand over time. For example, election officials can inform the public about the relatively small number of voters arrive at the polls in the early afternoon. Also, hospitals can work to try to schedule patients more evenly over the morning to avoid early rushes and their negative effects on waiting times and patient outcomes.

5. *Using "kanban cards"*: mixing production and eliminating batch operations will, by themselves, tend to reduce the amount of work in process (WIP) inventory. Yet, the TPS system uses a "kanban card" system to directly limit the amount of inventory. Each item in the inventory buffer is assigned a card. When a subsystem runs out of kanban cards, the upstream system must stop its production. The effects can ripple upstream until the entire upstream system is shut down. This is obviously inefficient in the short term. However, part of the lean dogma is that shutting down the line will cause resources to be focused on complete and permanent problem resolution.

The basic vision of the TPS is to deliver exactly what the customer orders quickly and with zero waste of any kind. Mass producers typically make items in large batches to reduce the cost of setting up machines for different types of parts. This choice causes many types of problems beyond the inventory carrying costs associated with storing batches. Some of these costs are difficult to simulate including the quality losses because when problems are discovered they can affect all the parts in large batches.

The TPS dogma can inspire many alternative systems to study with simulation. In general, discrete event simulation can help "mass producers" answer questions about how their operations would be different if they adopted the TPS. Specifically, if they purchased equipment to facilitate smaller batches, what would be the effects on WIP and lead times? How much shorter would set up times need to be for mixed production to have a negligible effect on throughput? In relation to election systems examples, officials could use simulation to tradeoff machine purchasing costs with costs associated with leveling demand, e.g., through direct mail.

7.3 Sample Project: Lessons from the 2008 Election in Ohio

This section documents a sample project from project definition through decision support. The project was performed in preparation for two Ohio election summits sponsored by then Secretary of State Jennifer Brunner. It is the latest in a series of projects. No support was provided by the state for this project. The purpose was to use simulation to shed light on the 2008 November election with the aim of helping officials in other states as well as Ohio understand what works in designing efficient voting systems and also helping those who would like to cut costs make sure they do not cut vital programs in the process.

7.3.1 Project Definition: Learning What Works

The 2008 presidential election was historic in its election of Barack Obama. The first theory inspired allocation application approach was in 2006, also in Franklin County, Ohio. Another, albeit far more minor, historical event was the application of what is apparently the second voting machine allocation inspired by simulation and queuing theory. We call this approach "utilization-based" allocation, which required only simple spread-sheet-based formulas. It therefore offered reproducibility and transparency. Instead of allocating machines based solely on the number of voters expected, the allocation was based on the expected amount of time voters at that location would need to utilize the machines. Utilization is proportional to (the number of registered) × (the average time required to vote at the specific location). This formula is simpler than the Kolesar and Green (1998) formula from Chap. 6 and appeared to foster competitive waiting line performance in our simulations prior to the actual election.

The utilization-based approach was applied in Franklin County and minimal waiting times generally resulted. The main purpose of this report is to clarify the positive effect of the utilization-based allocation and of other factors in recent election. The factors being studied include: the directives from the legislature and Secretary of State Brunner's office for early voting and voting using paper ballots on Election Day. Also, additional factors include the benefit of statistical simulation for forecasting waiting lines prior to elections and the utilization-based allocation.

The motivation for this clarification relates to problems being experienced elsewhere across the county. The Associated Press documented many long lines in several parts of the country and one article claimed that waiting lines were the most important election systems problem. The CNN hotline cited "poll access" as only the third most important issue, which could be attributable to the less common but more troubling phenomenon of being excluded from voting because of registration issues or worrying that, because of machine failures, votes would not be counted. Our purpose here is to argue that, by applying utilization-based allocation and simulation, voting systems in these other parts of the country can be improved with minimal cost to taxpayers.

The charter for the project is shown in Table 7.1. The charter clarifies that the project focuses only on Franklin County, Ohio, which contains the city of Columbus and includes 532 polling places with Election Day turnout of approximately a half a million voters.

In Franklin County, the majority of the voting is done using direct recording equipment (DRE) machines. By law, these machines create a paper reel that is technically the official ballot. The county is particularly interesting, perhaps, in that almost all of the major methods for voting were available to residents. These including voting by mail in the form of no-fault absentee voting, early voting in person, Election Day voting with DREs, or, by voter choice, paper ballots scanned by optical scanner. Here we focus only on waits for those using DREs and not including any effects from waiting at registration. Therefore, actual waits would

Table 7.1   Charter for project to derive the lessons from the 2008 election in Ohio project

Project	Lessons from the 2008 Election in Franklin County, Ohio
Team members	Theodore Allen and Mike Bernshteyn
Timing	The 3 weeks following the election
Direct costs	$0 (we volunteered our time to do this analysis)
Special responsibilities	Mike made the software available and Theodore ran the simulations and documented the results
Scope	The 532 voting locations in Franklin County, Ohio where direct recording equipment (DRE) machines are the primary method of voting. The allocations in each scenario were given based on historical facts
Primary objective	To clarify the effects on the expected latest poll closing time (the key response) of using HAVA compliant machines (factor A), using more machines (factor B), allowing early voting combined with a simulation-based warning (factor C), allowing same-day voting using paper ballots (factor D), and using the utilization-based machine allocation method (factor E)

likely be longer than the models predict. The effects of absentee voting and other options were simply to reduce the pool of voters for DREs.

This exotic situation was costly and a potentially important issue for the future is cost savings with minimal losses in terms of waiting and security. By studying the election after it occurred, we have extremely high quality data sources including timing data from our previous project. In the previous project, we performed a mock election and timed representative voters on actual machines using ballots of different lengths (Allen and Bernshteyn 2008).

7.3.2 Input Analysis and HAVA Compliance

This section, derived from our report in Allen and Bernshteyn (2008), describes the data collection or input analysis with regard to the DRE service times. The Help America Vote Act (HAVA) had many effects on US election systems. One of the least noticed effects was the dramatic increase in the time it takes for voters to cast their ballots using DRE. Partly because of the mandated warning related to under-votes and because of the implicit encouragement for non-full faced equipment, voting times in Franklin County increased dramatically as shown in the figure below. Non-full faced machines are generally like PCs which handicapped people can reach easily but which require the user to page through multiple screens to cast a ballot.

The figure shows the distribution of the times voters need to cast ballots once they are permitted to monopolize the voting machines. The first curve refers to non-HAVA compliant, full-faced machines (from 2004) and subsequent curves refer to the compliant ES&S DREs used in 2006 and subsequent elections. The average times roughly doubled and some voters needed more than 20 min in the recent elections (Fig. 7.1).

The result is that: (1) after 2004 substantially more time was needed to cast your ballot once you reached the DRE and (2) the variation from location to location also increased. Addressing both challenges without unnecessary expenses requires an attention to the fundamentals of waiting systems. This follows because elementary queuing theory teaches us that slowing the service times of a machine (by making using it take longer to cast ballots) has a similar effect to increasing the arrival rate of machine users (e.g., an increase in the number of registered voters). Officials in many parts of the country noticed this effect but did not know precisely how to predict its consequences.

7.3.3 Simulation and Non-Interacting Systems

Figure 7.2 illustrates the 532 polling locations. Each voter is assigned to a precinct in a location which is a type of "vote center" in election systems terminology.

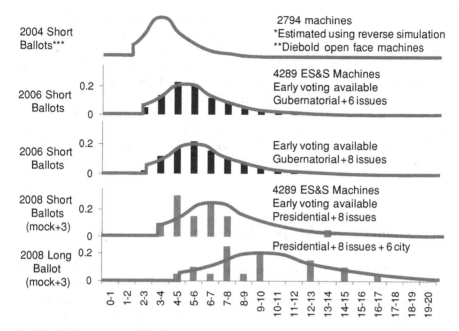

Fig. 7.1 Estimated voting time distributions (time it takes to vote when at the machine)

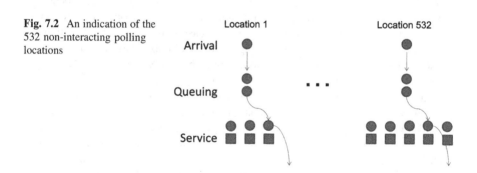

Fig. 7.2 An indication of the 532 non-interacting polling locations

The assignment is made in part because different precincts generally vote on different issues and races. Because of technology improvements, voters within their location can vote on any machine, i.e., machines are not dedicated to precincts. Different locations do not interact because their voters and resources are not shared. There are effectively 532 non-interacting systems.

Each system has a different arrival process because the number of registered voters is different (and the demographics are different). Each system has a different service process because the length and nature of the ballots are different. In 2008, all locations voted on 24 races. However, some locations voted on 19 ballot initiatives (estimated mean service time equal to 12.5 min) while others voted on 8 ballot initiatives (estimated service time equal to 6.5 min).

The key element in voting systems simulation is to recognize that the overall system performance is evaluated based on the expected waiting lines at the worst performing precinct. Therefore, each simulation replicate or run must include a day of voting at all 532 locations and aggregation based on the worst performing location. Simulating 0.5 million voters takes time and the large number of locations means that standard software might likely be too slow and unwieldy. As a result, my business partner, Mike Bernshteyn, of Sagata Ltd. developed C++ code to perform the needed simulation. Following Allen and Bernshteyn (2008), we used an empirical distribution for the non-homogenous arrival process and a Poisson breakdown process for the machines.

7.3.4 Output Analysis: Studying Alternative Scenarios

In our report to Franklin County, we applied simulation to predict that long lines could be expected if early voting did not greatly exceed the previous levels. In both the 2006 gubernatorial and the 2008 primary elections, approximately 24% of all ballots cast were via early or non-election day voting. Because of the longer ballots and HAVA compliance, we applied straightforward simulation and predicted lines comparable to 2004.

At the time of our report, Franklin County was already in the process of initiating a substantial advertising campaign encouraging people to vote early. This campaign was likely strengthened (to at least some extent) by our findings which were publicized widely including on the Columbus Dispatch front page. Whatever the cause, the result was that in November 2008, approximately 44% of the ballots were cast not on Election Day in Franklin County.

Our simulations here are generally based on voters arriving randomly over the day and machines also breaking down at random intervals. Further, we consider an overall turnout and 4 scenarios two being higher and two lower than the overall turnout fraction. Here, we focus on post election simulation with known turnout at each location. For simplicity, we use truncated normal voting times and Poisson arrivals spread uniformly over the 13 h Election Day. In Allen and Bernshteyn (2008), we had used empirical distributions (see Chap. 3) of voting time and turnout percentage distributions causing generally minimal differences in predictions.

The table below describes seven scenarios relating to the historical changes in the Franklin County election systems. Scenario 1 begins with what, hypothetically, could have happened had the 2,870 full-faced machines from 2004 simply been used in 2008 with its more variable ballot lengths. Subsequent scenarios describe the changes to HAVA complaint machines, the increase in the number of machines, the switch to early voting, and the advent of Election Day in-person voting using paper ballots. Scenario 6 describes the system used in the actual November election. Scenario 7 refers to a different hypothetical scenario in which only Election Day DRE voting was permitted and utilization-based allocation was applied (Table 7.2).

Table 7.2 Description of scenarios simulated in voting systems project

Change	Scenario						
	1	2	3	4	5	6	7
New HAVA compliant machines	No	Yes	Yes	Yes	Yes	Yes	Yes
Larger number of machines	No	No	Yes	Yes	Yes	Yes	Yes
Early voting and simulation warning	No	No	No	Yes	Yes	Yes	No
Election day in person on paper ballot voting	No	No	No	No	Yes	Yes	No
Utilization-based DRE allocation	No	No	No	No	No	Yes	Yes

The simulation results based on 10 replications show the minor negative effect on the poll closing times of the 2008 ballot lengths which, by themselves, would likely have made 2008 slightly worse than 2004. Next, shown is the effect of the HAVA compliant machines, which greatly worsen the waiting line situation. This very large, undesirable effect is countered by the next effect in scenario 3 of early voting for which the secretary of state and the legislature deserve much credit. As noted previously, we also claim some credit for helping to increase the early voter percentage from 24 to 44%. This change caused by far the largest and most desirable effect on the expected poll closing time and the expected waiting times in general.

Next, we considered the effect on Election Day voting of using paper ballots. We assumed that these voters simply waited through the line and went off to a different area than that used by DRE voters. Therefore, the effect from the waiting line point of view was as if the voters did not exist or a two line system was applied. Next, our simulations indicated that a statistically significant improvement is associated with utilization-based allocation. This improvement could only have been relatively small because the lines were already expected to be short as a result of the other improvements. The last scenario shows the relatively substantial effect of utilization-based allocation if early voting and Election Day paper voting had not been permitted. All results are shown in Fig. 7.3.

7.3.5 Decision Support

The results in Fig. 7.3 show that Ohio "dodged a bullet" because of the decisive actions of the legislature and Secretary of State Brunner. The HAVA compliant machines require such greater service times that simply doubling the number of them would likely not have been sufficient. The added machines alone would have resulted in expected actual poll closing times 15 h after the doors officially closed at 7:30 pm. Ohio would have been again the subject of national ridicule. Also, thousands of citizens would probably have been deterred from voting. Without utilization-based allocation, these deterred voters would likely have been concentrated in the downtown area of Columbus where the ballots were longest (slowest service times). The county would then be open to equal-access-based law

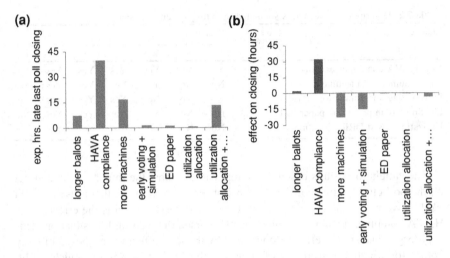

Fig. 7.3 Bar charts of the: **a** expected poll closing times and **b** factor effects on closing times

suits because the allocation would effectively discriminate against voters who have longer ballots among whom many are African Americans.

Fortunately, we did have decisive leadership. The early voting process removed 211,000 voters from the pool on Election Day. We like to feel that publicity given to our simulation analysis contributed to this success by helping to encourage many voters to use no fault absentee voting. This illustrates the potential power of simulation to anticipate and avoid problems before they happen.

Because of the more powerful actions by the secretary of state and the legislature, the effects of our utilization-based allocation were limited. However, other counties and states may not budget all the expenses needed for early voting in two forms (by mail and in person) as well as Election Day voting in two forms (DRE and paper). Also, future elections might see even higher levels of voter turnout. In these other cases, sensible allocations must account for both the arrival processes (numbers of registered voters) and the service processes (the potentially variable ballot lengths). Variable ballot lengths and service times are not an issue in many locations around the country. However, in the locations where they are an issue such as Ohio and California, it is obvious to any student of simulation and/or queuing theory that they should be accounted for.

To generate sensible allocation formulas, what is needed is the number of registered voters, a list of issues at each location, and estimates of the average shortest and longest average voting times. Then, using Excel, one can quickly find the target utilization ρ_0 that either meets performance goals which can relate to the available budget for machines (i.e., the goal is to hit the target available) or waiting time performance goals.

Our conclusion is that simulation should generally be applied to predict possible catastrophes and to help ensure that sufficient resources are in place to avoid them. Applying simulation software such as that available from Sagata Ltd. should

Initialize: $\rho_0 = 0.85$;
Repeat{
Repeat for all locations {

$$\text{number of machines} = \text{roundup}\left\{ \frac{(\text{avg. voting time mins.})(\#\text{registered})(\text{turnout})}{(\rho_0)(13\ \text{hr})(60\ \text{mins./hr.})} \right\} \quad (7.1)$$

If{ performance is too poor} increase ρ_0;
If{ performance is too good} decrease ρ_0;
Until{performance is acceptable}

then be used to confirm that the waiting line performance is acceptable. For each county, the process takes only a few hours once the timing data for the shortest and longest ballot styles are available.

Also, for voting machine allocation in counties with variable ballot lengths, some equation that includes the effects of both the turnout and the ballot lengths should be applied. Indeed, it would probably be desirable for federal law to make a theory inspired formula explicit as part of the existing guarantee of equal access to voting. Equation 7.1 is one candidate for such a formula as is the Kolesar and Green (1998) formula in (6.2).

7.4 Case Study of Supermarket Checkout Fast Lane

In this section, a project more similar to the scope of a student project is described. It is designed to provide an example answer to question 8 of Sect. 7.6. The model has limitations that are also discussed, but it is argued that the recommendations are likely reasonable within the confines of the defined scope.

Abstract: The application of discrete event simulation to support decisions about designating one cashier for customers with X or fewer items is investigated. Responses considered include average line length (Y_1) and expected store revenues (Y_2) under specific assumptions about losses as they vary with line length. The scope of the project and recommendations relates only to a rush period from 1 pm to 3 pm weekdays when all cashier aisles are routinely in use. The recommendation that $X = 5$ items for that period is described.

7.4.1 Problem Definition

An Ohio supermarket caters to gourmet customers such that they expect high service levels even during their rush period. This makes self-service options like the ones explored using simulation by Opara-Nadi (2005) irrelevant.

During all periods the store uses a cascading policy in which workers from other areas of the store are pulled into become cashiers quickly when lines reach certain levels. The use of simulation to inform decisions about such triggering policies in the context of retail checkout lanes is the subject of Williams et al. (2002).

The focus in this project is on a rush period after lunch in which the maximum number of aisles open is assumed to be three, an assumption based on informal observations of aisle service times during the 1–3 pm time period. During these time periods it is assumed that additional cashiers are not available. The key input factor or decision variable "X" explored is the maximum number of items allowed in a fast lane aisle. If that value is $X = \infty$, the there is effectively no dedicated fast lane aisle. The scope is described in Table 7.3.

The responses considered are the average waiting time (Y_1) and average revenue lost (Y_2). We assume that customers are lost if they observe all lines having six or more other customers. This is a gourmet food store and customers value their time highly. We have observed actual sales losses in cases in which all lines were too long. Further, we use a linear regression model to relate cashier service time to revenue as described in the input analysis.

It is currently not estimated (unknown) how much revenue is being lost because of excessively long times. However, one project goal is to estimate the current loss rate and to minimize the expected average loss per 2 h period (Y_2).

7.4.2 Input Analysis

The waiting line system involves three aisles each with its own queue. The system involves arrivals who appear in roughly equal numbers at aisles 1 and 3 and then select their queue by observing the queue lengths. The arrivals are clearly unco-ordinated. Often more than one person arrives at the same time but they typically share the purchased goods. From the modeling point-of-view, they are equivalent to individual arrivals and not batch arrivals.

We asked the managers for the historic average number of customers arriving during the 1–3 pm time period. Their estimate was 180 purchases leading to an estimated arrival rate of 0.4 arrivals per minute and exponentially distributed interarrival times. They told us that this estimate was a bit on the high side but of interest. We timed 20 service times by cashiers as well as the purchase revenues. The results are shown in Table 7.3 together with events that we noted.

Next, to model the service time distribution, a relative frequency histogram was constructed as described in Sect. 3.2. The results are shown in Table 7.4. The associated continuous distribution and the best fit two-parameter exponential distribution is shown in Fig. 7.4. Using the excel solver, the best fit two parameter exponential distribution is computed to be $0.21 + EXPO(0.388)$ in minutes and using the ARENA notation from Chaps. 10 and 11.

Next, we developed a simple model to estimate the revenues associated with each purchase. The scatter plot and linear regression model indicated in Fig. 7.5

Table 7.3 Charter for the gourmet supermarket fast lane study project

Project	Tuning the parameters of a supermarket fast lane
Team members	Theodore Allen and Andrew Allen
Timing	One day of data collection and modeling. One and one half days of analysis and documentation
Direct costs	$0 (we volunteered our time to do this analysis)
Special responsibilities	Andrew Allen ran the stopwatch. Theodore Allen was responsible for simulation modeling, analysis, and documentation
Scope	This project relates only to the gourmet food store with three aisles in operation in the weekday 1–3 pm periods
Primary objective	To clarify the policy for the fast lane aisle (X items or fewer) that is likely to foster desirable average waiting times and expected revenues under realistic long-term assumptions

Shopper	Time (s)	Purchase ($)	Notes
1	19.78	1.75	
2	46.89	43.27	
3	27.50	17.88	
4	30.42	14.16	
5	12.52	11.51	
6	18.21	11.40	
7	45.46	20.47	Customer is withdrawing cash
8	61.21	33.48	
9	9.40	7.19	
10	25.92	15.23	
11	18.64	7.20	
12	69.00	53.32	
13	46.76	17.00	Cash and call for new aisle opening
14	22.00	3.00	
15	52.00	47.19	
16	181.20	95.25	Customer is withdrawing cash
17	63.53	60.85	
18	39.40	5.30	
19	70.80	30.59	
20	12.50	5.10	

Table 7.4 Information for service time distribution histogram and fit

i	$q(i)$	Midpoint	Count	Bin	(Scale) rf(i)	$f(x)$	$\varepsilon(x)$
0	9.40	23.72	10	9.4–38.0	0.0175	0.0187	−0.0012
1	38.03	52.35	7	38.0–66.7	0.0122	0.0091	0.0031
2	66.67	80.98	2	66.7–95.3	0.0035	0.0045	−0.0010
3	95.30	109.62	0	95.3–123.9	0.0000	0.0022	−0.0022
4	123.93	138.25	0	123.9–152.6	0.0000	0.0011	−0.0011
5	152.57	152.57	1	152.6–181.2	0.0017	0.0007	0.0010
						SSE	0.000019

Fig. 7.4 The scaled relative
frequency histogram and the
two parameter exponential fit

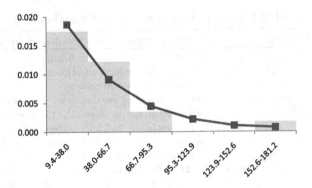

Fig. 7.5 Regression fit
showing the relationship
between service time and
revenue

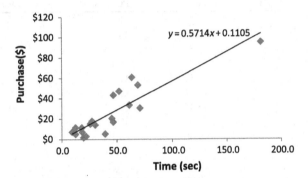

suggest that purchased revenues ($) can be predicted from service times, x (in minutes) using $0.5719x + 0.1105$. In our experience, we have observed customers giving up and leaving the store without purchasing items (reneging) whenever the shortest line is six or more customers. Therefore, to estimate the expected losses, we use the regression equation to predict the revenues that customers would have brought had they not left because of their simulated service time.

7.4.3 Simulation

The simulation model was built using ARENA software described in Chaps. 11 and 12. Figure 7.6 shows the ARENA model used starting with the "Create" entity block and stopping with the two "Dispose" blocks. We generated the service time first and used the "Assign" block to attach it to the entities as a variable. We did this because the financial revenue and line selection depends on the service times. The revenue is calculated using $0.5719x + 0.1105$ and a second "Assign" block, where x is the service time. Next, we checked that at least a single line had five customers or fewer. Otherwise, it was assumed that the customer walked away and their business lost.

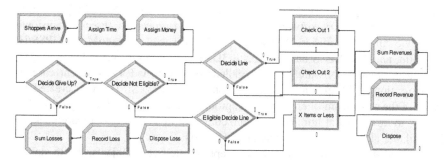

Fig. 7.6   ARENA simulation model used for the analysis

We made the further assumption that customers with smaller numbers of items (short service times) approach the cashiers on one side and shift over to the middle aisle if the line is shorter. Customers with large numbers of items (longer service times) approached the cashiers from the other side. This seemed reasonable because one side of the store has sandwiches and meal items and the other side has bulk items relevant to home sales.

The routing is based on the conditions of the queues. Specifically, the initial "Define" block is based on the following two way condition:

$$MN(NQ(X \text{ Items or Less.Queue}), NQ(\text{Check Out 1.Queue}),$$
$$NQ(\text{Check Out 2.Queue})) \leq CutOffLine \tag{7.2}$$

The second sets of "Define" blocks is also route-based:

$$NQ(X \text{ Items or Less.Queue}) \geq NQ(\text{Check Out 2.Queue}) \tag{7.3}$$

The loss accounting is based on record blocks that sum the losses or revenues of all entities reaching the entity.

Verification and validation was accomplished by having someone familiar with store operation walkthrough the simulations using the built-in ARENA animation feature. During these iterations, the lines were observed to be longer than usual but reasonable. As a result, the assumption of an average 180 customers per 2-h period is considered to be on the high side. The patterns of the lines and the high variation in the line length for the fast lane were considered realistic.

7.4.4 Modeling Using Alternative Software: SIMIO

This section is not part of the case study and merely shows how alternative software might offer additional benefits. In Chap. 10, it is argued that ARENA is used in university instruction mainly because of the short learning curve. However, it is probably true that most businesses use alternative software packages having superior animation or visualization capabilities. For example, ARENA is a mainly two-dimensional (2D) software with limited post-processed three dimensional (3D) viewing capabilities. Software packages such as AutoMod, GPSS/H, ModSim, and WITNESS have powerful 3D or virtual reality viewing.

One promising software package called SIMIO provides equivalent ease of learning along with enhanced visualization capabilities. SIMIO was created by much of the same team of engineers that produced ARENA. A trial copy of SIMIO is easy to download at http://www.simio.com and the complete software package is available at no charge for academic institutions.

Since SIMIO is built around the concept of Intelligent Objects, the modeling approach is more intuitive. It is based on drag-and-drop placement of objects, rather than drag-and-drop logical modules. Each of these objects represents a physical component in the real system. For example, to represent the same system described in Sect. 7.4.3, we would start by placing three Servers from the library on the left—each representing one of the checkout counters. Then we would place a "Source" and a "Sink" representing entry and departure locations and connect them with paths.

We change the properties of each object to represent intended behavior. For example, we set the "Source1" object properties to the expected arrival rates (exponential with a mean of 0.4 people per minute) and assign characteristics such as Service Time to the incoming shopper. As the entity leaves Source1, we would also select the server by specifying a selection goal and condition to select the least busy server appropriate to the customer purchase. This would result in the model shown in Fig. 7.7.

The difficulty of generating 3D models as well as the difficulty of drawing or obtaining appropriate 3D symbols have long been significant barriers to creating good animations. The model in Fig. 7.7 is shown in a top-down view with all default names and animation. It is shown in development and animation as a 3D model. In using the SIMIO program, simply pressing the "2" and "3" keys toggles the animation in real time between 2D and 3D.

We can replace the default symbols by using symbols from the included symbol library. For example, if we animate the checkout clerks, we simply select one of the many 3D people included in the library. We do the same to animate a variety of customers. Given pre-specified symbols (like a vendor-supplied symbol of the checkout counter), we could import them. We have yet another option to "download" symbols from Google 3D warehouse. This warehouse contains hundreds of thousands of 3D objects and SIMIO is the first discrete event simulation software that provides a direct link to the objects there. Searching "checkout" results in over 60 symbol options. Choosing among these options in addition to a few other quick enhancements results in the 3D animation shown in Fig. 7.8.

The animation in Fig. 7.8 is clearly more visually impressive and easier to understand than an animation like that shown in Fig. 7.6. Chapter 11 describes possible enhancements to ARENA animation but these are 2D. Real-time, interactive 3D visualization clearly provides a discernable advantage in verification and

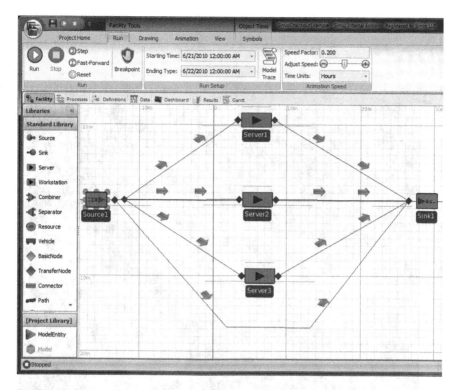

Fig. 7.7 SIMIO default simulation model of the checkout operation

validation. Perhaps most importantly it engages stakeholders emotionally, building trust and potentially aiding in the acceptance of the conclusions.

We run the model interactively using the "Run" menu both to see the 2D and 3D animation and to obtain preliminary results. But for experimentation, we run using the "Experiments" window. This allows us to set up and run multiple scenarios with multiple replications somewhat similar to the Rockwell Process Analyzer (Sect. 10.3). But SIMIO's experiments will also take advantage of all available processors to run multiple replications and scenarios concurrently (e.g. a quad processor will run four scenarios in about the same time as one). And the experiments can be linked to externally written add-ons to provide extended experimentation. One example of such an add-on is OptQuest, which provides sophisticated goal seeking algorithms to help in finding the best solution.

The output of the experimentation is expressed in multiple fashions. The experiment window itself displays values of key performance indicators (KPIs, or as SIMIO calls them Responses). These responses can also be displayed in Charts that provide extended box-and-whisker plots featuring MORE plot technology in Nelson (2008) to help accurately compare scenarios and identify the best.

While the experiment generates standard reports, many people prefer SIMIO's Pivot Tables. The full results are displayed in an interactive pivot table (Fig. 7.9),

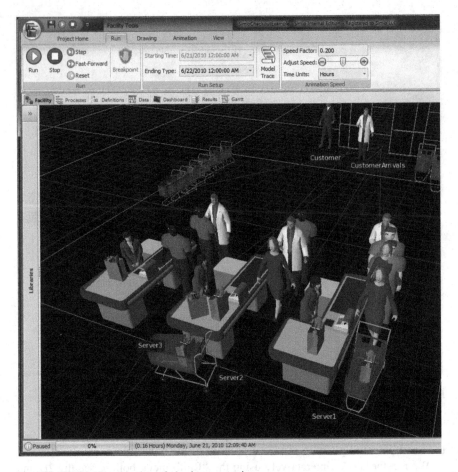

Fig. 7.8 SIMIO 3D animation of checkout operation

much like that of Excel. This output format is well suited to simulation analysis as it allows us to sort, filter, and categorize data thus making it easier to "data mine" the full value of the information buried therein.

7.4.5 Output Analysis

After verifying and validating the ARENA model in Fig. 7.4, 20 replicates were run for each of three levels of the factor of interest. The factor was the maximum service time that is allowed to be processed on the "X items or fewer" cashier aisle. We considered three levels of this factor: 0.25 min (15 s), 0.5 min (30 s), and 1.0 min (60 s). These times are assumed to correspond to the number X equal

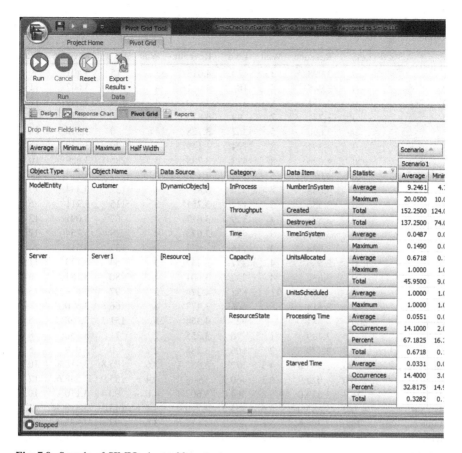

Fig. 7.9   Sample of SIMIO pivot table output

to 3, 6, and 12 items respectively. The simulation results based on 120 min of simulated time for the 20 replicates are shown in Fig. 7.5.

To achieve defendable confidence intervals, the twenty replicates were grouped into five batches of size four values. The simultaneous intervals are shown in Table 7.5 and Fig. 7.10. The results show that the sample size is sufficient to make a defensible recommendation.

7.4.6 Decision Support

The simultaneous intervals in Fig. 7.10 provide sufficient evidence for a recommendation. Under the modeling assumptions described in the input analysis section, the shortest service time cutoff (0.25 min) offers significantly reduced average waiting times compared with the middle setting cutoff (0.5 min)

Table 7.5 Results for three alternative systems that differ by cutoff time in minutes

Replication	0.25			0.5			1		
	Loss	Revenue	Wait	Loss	Revenue	Wait	Loss	Revenue	Wait
1	0	3,998	3	70	4,013	3	15	3,401	9
2	0	3,945	4	182	4,011	4	59	3,602	11
3	0	4,251	8	338	4,090	4	229	3,499	12
4	0	3,781	2	99	3,815	3	0	3,173	11
5	0	3,690	5	362	3,728	4	0	3,312	16
6	0	4,351	8	724	4,149	5	120	3,540	15
7	0	4,026	8	473	3,788	5	286	3,629	10
8	0	4,025	5	282	4,110	4	337	3,714	10
9	113	4,048	7	526	3,889	4	152	3,384	13
10	0	3,356	3	122	3,251	3	115	2,711	6
11	0	3,574	4	329	3,617	4	67	3,091	13
12	0	3,657	9	479	3,673	4	28	3,254	12
13	0	3,988	3	216	3,873	4	132	3,518	14
14	0	3,536	4	277	3,447	3	64	3,133	11
15	0	3,632	4	232	3,601	4	280	3,181	6
16	0	4,376	11	688	4,274	5	97	3,821	15
17	0	3,614	8	608	3,473	4	166	3,370	9
18	300	4,481	11	925	4,354	5	151	3,768	15
19	0	3,429	1	0	3,355	2	71	3,156	8
20	0	3,379	2	106	3,288	2	0	3,089	5
Batch 1	0.0	3,993.8	4.3	172.1	3,982.4	3.6	75.8	3,418.8	10.9
Batch 2	0.0	4,022.9	6.5	460.2	3,943.9	4.4	185.9	3,548.6	12.7
Batch 3	28.3	3,659.0	5.8	364.0	3,607.4	4.0	90.4	3,110.1	10.8
Batch 4	0.0	3,883.0	5.6	353.2	3,799.0	4.1	143.3	3,413.3	11.2
Batch 5	75.0	3,725.6	5.6	409.7	3,617.6	3.6	97.1	3,345.4	9.3
SD	32.8	160.8	0.8	109.0	175.9	0.3	45.4	161.5	1.2
Lower limit	−50.4	3,508.1	3.9	115.3	3,408.3	3.2	20.1	3,017.0	8.4
Est. Exp. Val.	20.7	3,856.8	5.6	351.8	3,790.1	4.0	118.5	3,367.2	11.0
Higher limit	91.7	4,205.6	7.3	588.4	4,171.8	4.7	216.9	3,717.5	13.6

and is likely shorter than the long cutoff setting (1.0 min). At the same time, the revenue loss is expected to be significantly less using the short cutoff (0.25 min) compared with the long cutoff (1.0 min). Other differences were not statistically significant.

Because of the benefits for expected waiting and lost revenue, we suggest that using the shortest cutoff is most desirable. We have argued that this corresponds to $X = 3$ items or fewer. Therefore, we recommend having a fast lane aisle during the 1–3 pm time periods and adjusting the maximum allowable number of items down from 10–3. It might also be advisable to put the proviso "no checks" on the sign since there is a suggestion that paying by check greatly increases the service times.

Limitations of the model include the problem that customers with long service times may be abandoning their items based on conditions in the fast lane, which

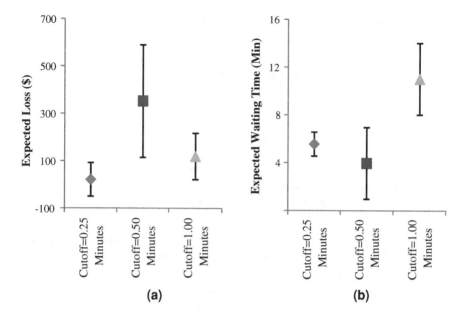

Fig. 7.10 Simultaneous intervals for **a** expected loss; **b** average waiting times

they would not be allowed to enter. Also, more data would be needed to evaluate the impacts of allowing payment by check or whether eliminating checking options would be annoying to customers and might therefore be unadvisable. Finally, studying the model during more realistic arrival conditions could be worthwhile. The assumption of an average 0.4 arrivals per minute resulted in lines that were longer than average observed during the verification and validation stage (but not unrealistic). Yet, it seems likely that the recommendations would also offer benefits under various arrival assumptions.

7.5 Project Planning Exercise

As described in Chap. 1, a critical step in each project is developing the charter. After describing each of the other steps in a simulation project, the reader may see more clearly the relationship between initial choices and later actions. For example, leaving specific subsystems out of the scope can mean that data on these subsystems is not needed in input analysis. Also, the decision to include an additional factor made in project planning can imply a need to build and evaluate several alternative systems in output analysis. This section describes an exercise that explores these possible inter-relationships. It also emphasizes the need for imagination in project planning and successful project execution.

Step 1. Pick one of the following systems to study:

     i. A book store
     ii. A dept. of motor vehicles drivers' license centers
     iii. A local Dominos delivery operator
     iv. A local McDonald's
     v. A local Starbucks
     vi. An emergency room
     vii. A local coffee shop you know
     viii. A restaurant
     ix. A university food service location
     x. A student health services clinic
     xi. A student services in Lincoln tower
     xii. A university operator

Step 2. Develop a team charter for a project which should focus on one specific subsystem and time period. Your charter should include factors whose settings might be changed and responses of potential interest. Use your responses to express the goal of your hypothetical project.

Step 3. Create a work flow or flow chart of your system based on your imagination. You will not have time or access to observe the real system. If you feel that there is a truly important aspect of uncertainty, create two or more work flows of the same system.

Step 4. Identify any rare events that might have a truly important impact on your system responses. Estimate the chance that these events will happen for any given entity in your system. For example, assume there is roughly a $p_0 = 0.01$ probability that an arbitrarily selected voter would be handicapped.

Step 5. For every process in your flowchart(s) and the arrival process, describe how much data you would collect and during which time period. Make sure to match your data collection strategy with the goals in your charter and the possibility of rare events.

Step 6. (Optional) Guess which distribution might be a logical choice for interarrivals and for each service distribution.

Step 7. Document your findings in a 1 page handwritten summary.

7.6 Problems

1. Describe the possible application of the Theory of Constraints to a simulation project involving a popular coffee house. In your answer, clarify the role of the theory in the project.
2. Describe the possible application of the Theory of Constraints to a simulation project involving a call center. In your answer, clarify the role of the theory in the project.

3. Describe the lean production way of making five toasted peanut butter and jelly sandwiches on a family picnic.
4. Describe the lean production way of grading exams. Make reference to the current way which involves grading each question 1, then each question 2, then each question 3, and so on.
5. List some possible ways election officials can level demand in election systems.
6. List some possible ways that universities can level demand for teaching services.
7. How can simulation help in making decisions about kanban cards?
8. (Class project)

   Team: You can perform this project individually or with up to 3 people. However, if you include more team members, more is expected in terms of data collection and the relevance of results to real business decision-making.
   Report: Your report must have a 200 or fewer word executive summary or abstract and sections with titles similar or identical to:

   • Problem definition,
   • Input analysis,
   • Simulation,
   • Output analysis, and
   • Decision support.

   Tables and figures: These must be included in the key data, findings, and the modeling assumptions with captions and references to each figure in the text.
   References: Relevant references should be included and mentioned in the text. For example, if you are pursuing a model of a hospital room, you might write about how your model is different from the one in Medina et al. (2008) with the following reference included at the end of your report:

   Medina R, Vazquez A, Juarez HA, Gonzalez RA (2008) Mexican Public Hospitals: a model for improving emergency room waiting times. In: Mason S, Hill R, Mönch L, Rose O (eds) Proceedings of the 2008 winter simulation conference

   Requirements: The developed decision-support must have hypothetical relevance to some imagined decision-maker. The more people are on the team, the more relevance is needed. The data collected in the input analysis must be 100% real with a brief description of the details of the data collection including the equipment used. Any model should have at least three of the following five characteristics:

   (a) Real world data used for fitting all distributions,
   (b) Two or more alternative scenarios studied with simultaneous intervals, and

(c) At least one alternative scenario of plausible interest.
Additional points may be earned for using:

(a) Resource sets, i.e., individuals who can provide types of service,
(b) Station/routes, i.e., the inclusion of issues relating to material handling,
(c) Schedules (capacity, arrival, or both),
(d) The winter simulation conference author kit formatting, and
(e) Any other advanced modeling techniques.

Grading: 40% report quality (10% proper figure uses, 10% proper referencing, 10% writing style, 10% abstract or executive summary), 20% input analysis (10% data quality and 10% input analysis), 20% simulation/modeling quality, 20% decision support quality (10% output analysis and 10% correctness of conclusions).

Chapter 8
Variance Reduction Techniques and Quasi-Monte Carlo

Computer speeds continue to increase. At the same time, the complexity and realism of simulations also continues to increase. For example, 20 replicates of the voting systems simulation in Chap. 7 involve approximately 10 million simulated voters. Currently, a standard PC requires several minutes to yield the expected worst precinct closing time estimate. In 5 years, the identical simulation might require less than a single minute. Yet, we might choose to include in the simulation details about registration process and voter demographics so that the resulting time might require more than 10 min again.

Also, the voting simulation is small-scale compared with agent-based brigade-level war game simulation or discrete event simulations of large emergency rooms. Waiting for computers and Monte Carlo to estimate the expected waiting times or other outputs may require days or even months. Often, decision-makers are not prepared to wait longer than a few hours or overnight. Part of the challenge is the need for iteration of assumption parameters, i.e. the analysts discover the run results must be rerun because of mistakes in the input parameters. Probably the most common approach to reduce computing times is to simplify models by adding assumptions, e.g. registration has no effect on waiting times. These added assumptions can cause inaccuracies and losses of confidence in the results.

In this chapter, techniques designed to achieve greater computational efficiency than Monte Carlo simulation are described. Ideally, these methods can yield the desired estimates with improved accuracy and reduced computing time. This can permit greater detail in the simulations and/or permit the speeds necessary for real-time decision-making.

As noted in Chap. 2, the central focus of discrete event simulation theory is the evaluation of expected values. Also, expected values are integrals and a main method to estimate the values of integrals in Monte Carlo simulation. Chapter 3 described how Monte Carlo is accomplished in discrete event simulation using simulation controllers. This chapter focuses on alternatives to Monte Carlo estimation offering (potentially) reduced computing time to achieve estimates of

T. T. Allen, *Introduction to Discrete Event Simulation and Agent-based Modeling*, DOI: 10.1007/978-0-85729-139-4_8, © Springer-Verlag London Limited 2011

expected values with improved accuracy. Virtually all of the methods described here are so-called "variance-reduction techniques" and are documented in Wikipedia or in the archives of the Winter Simulation Conference.

It should be noted, however, that Monte Carlo is still a viable method. Yes, other approaches can achieve more accurate expected value estimates in reduced computing time. Yet, Monte Carlo and the batching for normality procedure described in Chap. 4, combined with the intervals in Chap. 5 provide a high level of defensibility. Also, Monte Carlo estimation is relatively simple. Simplicity implies fewer opportunities for systematic errors and greater transparency, i.e. users understand with improved intuition what the simulation is doing.

In Sect. 8.2, an overview of variance-reduction techniques (VRT) is presented together with quasi-Monte Carlo methods. Also, the method of common random numbers (CRN) for system comparison is presented. Section 8.3 describes methods based on inserting alternative sequences in place of pseudorandom U[0,1] numbers in discrete event simulation. These methods include Latin hypercubes, descriptive sampling, and quasi-Monte Carlo. In Sect. 8.4, a brief description of importance sampling is provided. Section 8.5 briefly describes techniques designed to derive more than a single number from a simulation run or replicate.

8.1 Variance-Reduction Techniques and Common Random Numbers

The phrase "variance-reduction techniques" (VRTs) refers to a collection of techniques intended to permit more accurate estimates than the Monte Carlo estimates described in Chap. 2. Some of these methods such as descriptive sampling derive "biased" estimates. This means that, if given infinite numbers of replicates, the resulting estimates would fail to converge to the true integral values. This is not true for Monte Carlo because of the central limit theorem. Yet, we might apply descriptive sampling instead of Monte Carlo because, for a small number of available replicates, the bias is small compared with the remaining so-called "variance" error (total error = bias + variance). Descriptive sampling would generally be expected to yield a more accurate estimate.

The list of variance-reduction techniques is long. These techniques include:

- Common random numbers (CRNs). CRNs use streams of pseudorandom numbers as in ordinary Monte Carlo. They focus on the comparison of alternative systems as in the output analyses in Chap. 5. By using the same random streams to evaluate alternative systems, they reduce the variance in the estimated performance differences.
- Latin hypercube (LHC) sampling. LHC sampling is a special case of orthogonal array LHC sampling (Tang 1993). These techniques offer no bias in estimation and proven improvements in the accuracy of the derived estimates.

- Descriptive sampling (DS). DS methods generate biased estimates. Yet, in numerical examples they appear to dominate LHC accuracy. Also, their application is slightly easier.
- Importance sampling. Importance sampling is critical for the study of high consequence but rare events. By artificially increasing their frequency and then weighting outputs, these methods can result in unbiased and reduced variance estimates of average system performance.
- Stratified sampling. For cases in which sub-populations or strata differ greatly, it can be of interest to sample each sub-population (stratum) independently. For example, in selecting voters for our mock election, we selected exactly 6 first time voters and 14 experienced voters to mirror the overall statistics of our population. Within each group we picked randomly.

Quasi-Monte Carlo is generally not considered a variance-reduction technique. Yet, these techniques described in Sect. 8.2 function in a manner similar to pseudorandom numbers in ordinary Monte Carlo. Also, they are naturally comparable to Latin hypercube sampling and descriptive sampling. Perhaps the main reason why they are excluded from the set of VRTs relates to their origin in the number theory branch of mathematics. Other techniques were developed by researchers in operations research, statistics, or combinatorics.

In the remainder of this chapter, all of the abovementioned variance-reduction techniques are described together with quasi-Monte Carlo with one exception. The exception is stratified sampling. We did apply stratified sampling in our election project, but not as a variance-reduction technique to speed-up our simulations. We used it for human subject selection in input analysis. This is probably typical of many discrete event simulation investigations.

8.1.1 Common Random Numbers

As noted previously, common random numbers (CRN) relate to comparing alternative systems and not to deriving more accurate estimates of individual systems. CRNs are important because they have no side-effects (like estimation bias) except for the minor difficulty of starting each system with the same seed. They are the subject of constant and continuing interest including extensions such as in Ehrlichman and Henderson (2008). Miller and Bauer (1997) showed how they can be built into subset selection and indifference zone techniques like those described in Chap. 5.

To illustrate common random numbers, we return to the registration and voting time simulation from Chap. 2. In this problem, simulation is unneeded because it is reasonably easy to directly apply calculus to estimate the expected summed time. Imagine that we are comparing an alternative direct recording equipment (DRE) machine that is associated with less variable voting times because of a wider display. We are interested in whether switching to these machines will reduce the key response which is the expected sum of registration and voting times. The two alternative systems are:

System #1 (S1):   Using the current equipment, we have TRIA(0, 0.229, 2.29) minute registration times and TRIA(4.0, 5.2, 16.0) minute voting times

System #2 (S2):   Using the new DRE machines, we have TRIA(0, 0.229, 2.29) minute registration times and TRIA(5, 6, 10) minute voting times

From calculus, we know that the expected time for system #1 is (0.0 + 0.229 + 2.29)/3 + (4.0 + 5.2 + 16.0) = 9.23 min. This follows from the formula for the expected value of triangularly distributed random variables as described in Chap. 2 in Eq. 2.11. Similarly, for system #2 we have (0.0 + 0.229 + 2.29)/3 + (5.0 + 6.0 + 10.0) = 7.84 min. Therefore, we know that system #2 is an improvement relating to the chosen mean response. However, suppose that we needed to derive this using simulation alone. Table 8.1 shows the simulation of system #2 (S1) using the same stream as that applied in Table 2.4 to simulate system #1.

The theory behind common random numbers is simple. Suppose that we intended to simulate the differences between the systems rather than the individual system values. Then, the differences in Table 8.1 would have similar properties to the individual system values, i.e. it is defensible to treat them as approximately independent, identically distributed (IID) random variables. Creating confidence intervals on the differences and checking that this interval does not include zero is equivalent to paired t testing. For example, the 95% confidence interval on the differences in Table 8.1 is 1.93 ± 1.39 min. This proves defensibly that the new system (S2) offers improved performance. This assumes that the differences are approximately normally distributed. Fortunately, if there is a concern about normality, the differences can be batched as for individual replicate values and described in Chap. 4.

Table 8.1 System #2 (S2) comparison with S1 using common random numbers (CRNs)

I	Z_i	U_i	Replicate	Registration	S2 Voting	S2 Time	S1 Time	Difference
0	19	–	–	–	–	–	–	–
1	44	0.698413	1	1.097	–	–	–	–
2	27	0.428571	1	–	6.619	7.716	8.491	0.775
3	31	0.492063	2	0.742	–	–	–	–
4	56	0.888889	2	–	8.509	9.251	12.947	3.696
5	39	0.619048	3	0.949	–	–	–	–
6	43	0.682540	3	–	7.480	8.429	10.535	2.105
7	5	0.079365	4	0.204	–	–	–	–
8	51	0.809524	4	–	8.048	8.252	11.236	2.983
9	55	0.873016	5	1.516	–	–	–	–
10	17	0.269841	5	–	6.179	7.694	7.788	0.094

8.2 "Better" than Pseudorandom

Many variance-reduction techniques (VRTs) are based on substituting alternative sequences in place of the pseudorandom U[0,1] numbers in Monte Carlo. Therefore, the process of transforming to specific distributions, applying simulation controllers, and aggregation described in Chap. 4 are retained. For example, in Table 8.1 we would simply replace the column labeled "U_i" with something other than pseudorandom numbers. Techniques of this type include quasi-Monte Carlo, Latin hypercube sampling, and descriptive sampling.

Therefore, all of these methods retain the basic Monte Carlo estimation structure and estimates derive from:

$$\text{Estimated expected value} \equiv \text{Xbar} \equiv (1/n) \sum_{i=1,\dots,n} X_i \qquad (8.1)$$

where the $X_1,\dots,X_n$ derive from processes that resemble Monte Carlo discrete event simulation.

8.2.1 Latin Hypercube Sampling

McKay et al. (1979) invented Latin hypercube (LHC) sampling to facilitate efficient integral estimation. Interestingly, in their subsequent work they focused on applications to computationally expensive computer simulation that did not involve random variables. Yet, their methods and extensions in Tang (1993), Lemieux and L'Ecuyer (2000), and others offer perhaps the most defensible alternatives to Monte Carlo simulation. If simulation runs are extremely computationally expensive, it is the opinion of the author that orthogonal array Latin hypercubes offer significant advantages for discrete event simulation.

Here, we describe only the basic method for LHC sampling. This is a special case of the more powerful but also more complicated methods in Tang (1993), which are based on so-called orthogonal arrays of strength two and higher. LHC sampling itself requires a sequence of pseudorandom numbers. Here, we apply the same sequence of numbers from the linear congruential generator from Chap. 2. In real applications, higher quality pseudorandom numbers would be applied.

Assume that we know in advance that we will be performing n replicates or simulation runs. For example, in the election systems example in Table 8.1, we might plan in advance on $n = 5$ simulations. In this example, performing $n = 10,000$ simulations is trivial in Microsoft® Excel. However, we will pretend that five runs is a significant computational challenge.

Step 1. For each of the m random variables in each simulation, create a permutation of the numbers 1, 2,..., n in a way such that all permutations are equally likely. Without loss of generality, we can use the sequence 1, 2,...,n with no re-ordering for the first variable. The first two columns of Table 8.2 show example permutations ($P_{i,1}$ and $P_{i,2}$) for the $m = 2$ variables (registration time and voting time). In practice, generating permutations itself uses pseudorandom numbers. We can generate the numbers 1, 2,...,n in one column and pseudorandom U[0,1] numbers in the next column. Sorting the first column by the second generates a random permutation

Step 2. Generate additional pseudorandom numbers (PRNs) for each run and variable, $U_1,...,U_{m \times n}$. Examples are shown in the third and forth column of Table 8.2, which are the linear congruential generator deviates used in previous examples

Step 3. Generate the Latin hypercube (LHC) deviates using:

$$\text{LHC } U_{i,j} = \left(P_{i,j} - 1 + U_{[j+(m)(i--1)]}\right)/(n) \quad \text{for } j = 1,...,m \quad \text{and} \quad i = 1,...,n. \tag{8.2}$$

For example, the fifth and six columns of Table 8.2 contain LHC deviates. These are to be used like pseudorandom numbers in Monte Carlo. In the example, we apply the inverse cumulative triangular distributions to transform them to generate simulated registration and voting times. The sample average of the resulting sums is 9.56 min, which is closer to the true value of 9.23 min compared with the Monte Carlo estimate of 10.2 min. Equation 5.2 predicts that 382 Monte Carlo runs would likely be needed to achieve a half width equal to this error, i.e. $h_0 = 0.34$. This shows the power of Latin hypercube sampling, which generally improves as more dimensions are involved and higher strength orthogonal arrays are used.

However, the half width from the LHC replicates is 3.0 min, which does not accurately reflect the true errors. This is also expected because it is not reasonable to assume that the sums 9.8, 7.5,...,9.5 are IID random numbers. They follow a rigid structure. More generally, LHC is statistically expected to yield more accurate estimates with the same number of total replicates. Yet, error estimation remains a challenge. Also, two pseudorandom numbers are consumed to generate a single LHC U_i deviate.

8.2.2 Descriptive Sampling

Saliby (1997) proposed a simplification of LHC sampling called descriptive sampling. It is based on a simplification of the LHC generation process involving only two steps.

Table 8.2 Simulation for estimating the S1 expected registration plus waiting time using a LHC

Perm. #1 $(P_{i,1})$	Perm. #2 $(P_{i,2})$	PRN U_i	PRN U_i (cont.)	LHC $U_{i,1}$ var. 1	LHC $U_{i,2}$ var. 2	Reg. time	Voting time	Sum (Min.)
1	4	0.698	0.429	0.140	0.686	0.275	9.618	9.893
2	2	0.492	0.889	0.298	0.378	0.470	7.020	7.490
3	1	0.619	0.683	0.524	0.137	0.791	5.421	6.212
4	5	0.079	0.810	0.616	0.962	0.944	13.778	14.722
5	3	0.873	0.270	0.975	0.454	1.944	7.588	9.532

Table 8.3 Simulation for the S1 expected registration plus waiting time using DS numbers

Perm. #1 $(P_{i,1})$	Perm. #2 $(P_{i,2})$	DS $U_{i,1}$ var. 1	DS $U_{i,2}$ var. 2	Registration time	Voting time	Sum (min.)
1	4	0.1000	0.7000	0.229	9.765	9.994
2	2	0.3000	0.3000	0.472	6.475	6.948
3	1	0.5000	0.1000	0.754	5.200	5.954
4	5	0.7000	0.9000	1.100	12.400	13.500
5	3	0.9000	0.5000	1.603	7.950	9.553

Step 1. For each of the m random variables, create a permutation of the numbers 1, 2, ..., n in a way such that all permutations are equally likely. (This is the same as for LHC sampling.)

Step 2. Generate the Latin hypercube (LHC) deviates using:

$$LHCU_{i,j} = (P_{i,j} - 1 + 0.5)/(n) \quad \text{for } j = 1,\ldots,m \quad \text{and} \quad i = 1,\ldots,n. \quad (8.3)$$

Table 8.3 shows the process of descriptive sampling for the expected registration-plus-voting-time estimation problem based on system #1. Averaging the simulated sums gives 9.19 min which is relatively close to the true mean of 9.22 min. Yet, again the estimated half width is 2.7, which is misleading. The true error is much smaller.

According to the sample size estimate in Eq. 5.2, Monte Carlo would require $n = 27,364$ to achieve an answer within $h_0 = 0.04$ of the true value. This assumes that we would want to set the half-width equal to the true error and we could somehow know the true error, i.e. $h_0 = 0.04$. The difference between $n = 5$ runs and $n = 27,364$ runs could be important for computationally intensive simulations. This shows the power of descriptive sampling.

8.2.3 Quasi-Monte Carlo Sampling

Quasi-Monte Carlo is relevant for at least two reasons. First, unlike LHC and DS sampling, quasi-Monte Carlo does not require the declaration of the number of replicates before simulation begins. As in Monte Carlo, the simulation can be terminated with little (if any) biasing in estimates whenever the decision-maker needs an estimate. Second, as with orthogonal array Latin hypercubes in Tang (1993) rigorous bounds are available relating to the expected accuracy advantages of the methods compared with Monte Carlo based on pseudorandom numbers. These results are described in, e.g. Caflisch (1998).

Here, we consider only a single type of quasi-Monte Carlo samples called "Halton" samples. These are based on a simple principle. After the first few points, the remaining points fill in the gaps, i.e. they are chosen to be as far apart from the other points as possible. Generally, the points 0.0 and 1.0 are assumed to be included in the existing points.

Table 8.4 (a) Base 2 and base 3 sequences, (b) simulation for the voting system using quasi-Monte Carlo

n	Base 2	Base 3			
(a)					
1	0.5	0.3333			
2	0.25	0.6667			
3	0.75	0.1111			
4	0.125	0.4444			
5	0.625	0.7778			
6	0.375	0.2222			
7	0.875	0.5556			
8	0.0625	0.8889			
9	0.5625	0.037			
10	0.3125	0.3704			
11	0.8125	0.7037			
12	0.1875	0.1481			
13	0.6875	0.4815			
14	0.4375	0.8148			
15	0.9375	0.2593			
N	Dim. 1	Dim. 2	Registration	Voting	Simulated time
(b)					
16	0.0313	0.5926	0.128	8.734	8.862
17	0.5313	0.9259	0.803	12.902	13.704
18	0.2813	0.0741	0.448	5.033	5.481
19	0.7813	0.4074	1.274	7.236	8.51
20	0.1563	0.7407	0.294	10.203	10.498
				Average	9.411
				Std. Dev.	3.007

Table 8.4a shows the so-called "base 2" sequence which puts the first point in the middle of the 0–1 range at 0.50. Next, it moves to 0.25 and 0.5 which are clearly both as far away from 0.0, 0.5, and 1.0 as possible. The sequence progresses until the pattern appears superficially like U[0,1] random numbers in its pattern but with less clustering. Therefore, our simulation starts with the 16th number in the sequence. For the second dimension, we simply change the initial number to be 0.33333 instead of 0.50. Next, we continue by again filling the gaps. Note that this approach might be considered greedy or myopic in the following sense. For any fixed number of samples, the Hamilton samples are not maximum-minimum distance or space-filling designs.

The simulation in Table 8.4b is again based on plugging the seemingly U[0,1] random numbers into the inverse cumulative triangular distributions. The quasi-Monte Carlo estimate is 9.41 min for the expected sum of the registration and voting times. Again, we know the true value is 9.23 min and the value represents an improvement in accuracy compared with our Monte Carlo simulation. The sample size formula in Eq. 5.2 suggests that $n = 1,355$ Monte Carlo samples would be needed to achieve this accuracy.

This assumes that we would want to set the half width equal to the true error and we could somehow know the true error, i.e. $h_0 = 0.18$. Comparing this with $n = 5$ samples shows the power of quasi-Monte Carlo. The half width is 2.7 min, which is again misleadingly high. The true error is much smaller.

8.2.4 Comparison of Alternative Techniques

Next, we summarize and compare the results from the voting systems numerical example. The qualitative aspects of the comparison here are factual. Yet, the quantitative comparison is based only on a single numerical example. These results are intended only to build rough intuition into the power of the alternative simulation methods. Results might vary greatly for different examples. Table 8.5 summarizes the results from the applications of the alternative methods to the same problem. In that example, the Monte Carlo (MC) was 10.20 min and the true value was 9.23 min.

This MC estimate was "lucky" in the sense that the sample size formula in Eq. 5.2 suggests that $n = 50$ samples would be needed to achieve a half width equal to 0.97. The other sample sizes are the number of MC samples (based on ordinary pseudorandom numbers) to achieve half widths equal to the actual errors. The results suggest that variance-reduction techniques such as descriptive sampling (DS) can achieve greater accuracy than ordinary Monte Carlo using hundreds of times fewer samples.

The ratios are simply the expected number of MC samples needed to make the half width equal to the error divided by five. This follows because the variance-reduction techniques achieved their accuracies with five samples. Also, as the number of dimensions and samples increases from two and five, certain methods, including orthogonal array Latin hypercube methods from Tang (1993), are expected to potentially improve in relative efficiencies.

To gain insight into the functioning of the various methods it can be helpful to inspect the geometry of the derived samples. Figure 8.1 shows the samples of numbers from the alternative methods in the $(0.1)^2$ square. These are the values that are then transformed into registration and voting times using the inverse cumulative triangular distributions. With only five pseudorandom numbers we see that the results fail to spread evenly over the square.

Table 8.5 Summary of absolute errors for voting systems case study and relative efficiencies

Method	Absolute error	Expected MC sample size (n for $h_0 =$ Error)	Ratio
Monte Carlo (MC)	0.97	50	10.0
Latin hypercube (LHC) sampling	0.34	382	76.4
DS	0.04	27,364	5,472.8
QMC	0.18	1355	271.0

Fig. 8.1 Scatter plot of five samples from the alternative methods

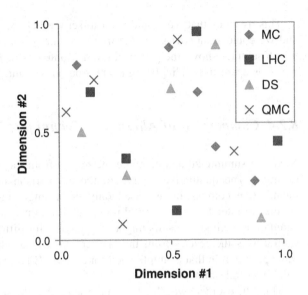

Table 8.6 Qualitative properties of the alternative methods to MC estimation

Technique	Variable n	Unbiased	Error estimation
Monte Carlo (MC, pseudorandom numbers)	Yes	Yes	Easy
Common Random Numbers (CRNs)	Yes	Yes	Easy
Latin Hypercube (LHC)	No	Yes	Difficult
Descriptive Sampling (DS)	No	No	Difficult
Quasi-Monte Carlo (QMC)	Yes	Some cases	Difficult

For example, there are no MC samples in the lower left quadrant. In comparison, the Latin hypercube (LHC) samples, descriptive sampling (DS) samples, and the quasi-Monte Carlo (QMC) samples are relatively spread out. By spreading out the samples, we are attempting to ensure that the wide variety of possible occurrences is experienced in our limited sample.

Table 8.6 summarizes the qualitative aspects of the alternative methods. "Variable n" refers to the possibility of terminating the method to apply an arbitrary sample size and derive defensible results. Monte Carlo simulation generates estimates with easily characterized properties for any number of samples. By contrast, if one were to stop simulation after performing half of the samples from an LHC array, a high degree of bias in the resulting estimates is almost guaranteed.

The "unbiased" column refers to the properties of estimates as the samples sizes approach infinity. Certain methods such as ordinary Monte Carlo are associated with zero bias errors. For example, with samples sizes in excess of 10,000,000, Monte Carlo would derive an estimate equal to 9.23 min in our voting systems example with probability near 100%. This is not true for descriptive sampling and certain types of quasi-Monte Carlo sampling.

The error estimation column refers to the ease of estimating the errors of the derived estimates. Under the assumption of approximately normally distributed simulated values, half widths based on Monte Carlo samples accurately characterize the errors of the estimates. If non-normality is detected, then batching can be applied for defensibility. Yet, standard confidence intervals approaches are not even approximately relevant for characterizing outputs of LHC, DS, or QMC simulations. The outputs are not identically distributed nor independently distributed to any good approximation. Therefore, while the accuracy of the derived estimates is likely to be improved, estimating the accuracy is difficult. The author believes that research into error estimation for variance-reduction techniques is an important subject which has received relatively little attention.

Note that the properties of common random number (CRN) methods are all desirable. The main downside is the potentially minor complication of using the same streams for all alternative systems being compared. Also, for complicated simulations uses large and potentially variable numbers of pseudorandom numbers, the benefits of common random numbers (CRNs) might be minimal.

8.3 Importance Sampling and Rare Events

Some systems are influenced greatly by rare events such as major machine breakdowns. If these rare events have high consequences for the systems of interest, then it is likely necessary to apply a technique called "importance sampling" to estimate the expected system response values. Importance sampling uses a custom selected distribution that generally makes the rare event more likely and then converts back to the distribution of interest. Here, we only comment on importance sampling and refer the reader to more focused textbooks on the topic, e.g. Srinivasan (2002).

In election systems, vote centers or locations are generally required to have at least three machines. As a result, the breakdown of a single machine is usually not associated with extreme consequences for the overall system properties. Also, if one includes paper replacement, machine breakdowns in the context of direct recording equipment (DRE) are not particularly rare. In our simulations, we typically assume that breakdowns arrive according to a Poisson process with mean interarrival times of 10 h.

8.4 Getting More Out of a Stream than the Batch Average

Deriving more than a single number from a stream of random numbers is generally considered critical for modeling the steady-state or long run properties of complicated systems. Often, in these systems there is an initial transient period for which results are not representative of the long run. The initial period is called the

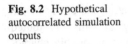

Fig. 8.2 Hypothetical autocorrelated simulation outputs

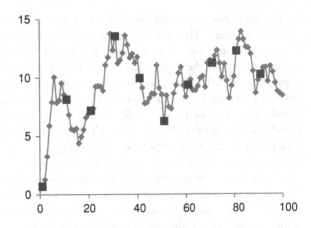

"warm up" and results from the warm up are discarded. For example, the system might start with zero parts in inventory even though the manufacturing plant never has zero work in process inventory. Therefore, the discrete event simulation would like to pay the computational expense of deriving realistic starting systems as small a number of times as possible.

In Chap. 4, the operation of discrete event simulation controllers is described. These controllers generate a series of event times and waiting times. The derived waiting times or other quantities in the series are generally neither independent nor identically distributed to any good approximation. Figure 8.2 shows the outputs from a discrete event simulation controller in a hypothetical example. These could be waiting times in minutes. Notice the initial warm up period in which the output quantity is near zero and then stabilizes. Consider also the fact that each output is relatively close to the preceding and following output. This phenomenon is referred to as "autocorrelation" and it makes the assumption of independently distributed random variables far less plausible. The sample correlation between successive points in this series is 0.89 indicating strong autocorrelation.

However, consider the outputs from every 10th simulation. Ignoring the initial output from the warm up period, the remaining values seem approximately independent, identically distributed. The sample correlation between each 10th observation and the following observation is only 0.21. This indicates small but potentially negligible autocorrelation. This situation suggests the following general framework which is implemented in many professional software packages:

Step 1. Discard simulation outputs from the warm up period (until the outputs approximately stabilize).
Step 2. Store every qth observation where q is chosen to be the smallest number such that the sample correlation between the qth observations is less than a cutoff value, e.g. 0.25.
Step 3. Analyze the stored observations in a manner similar to ordinary replicates.

Table 8.7 Toy example illustrating an application of the framework for generating semi-replicates

Voter	Reg. queue time (min)	Vote queue time (min)	Sum	Action	Reason
Person 1 (P1)	0.00	0.00	0.00	Discard	Warm up
Person 1 (P2)	0.00	6.49	6.49	Store	Done warm up
Person 1 (P3)	0.00	13.49	13.49	Discard	Skipping
Person 1 (P4)	0.00	14.12	14.12	Store	Next semi-replicate
		Average	10.3		

The stored outputs have sometimes been called "semi-replicates" and the sample correlation can be estimated using the "=CORREL()" function in Microsoft® Excel.

As a toy illustration, consider the results from the controller example in Chap. 4 shown in Table 8.7. The example is a toy because only 4 simulation outputs are available. As a result, there is no opportunity to evaluate whether the warm up period is over. Also, there is no chance to assess the extent of autocorrelation. The example is included simply to tie the framework to a case in which the relevant controller operation is described in detail.

For a more realistic example, we return to the data used to generate Figure 8.2. Table 8.8 shows only the first 30 outputs. Next, it shows the $q = 5$ and $q = 10$ sequences. Also, it shows the lags of these sequences. Lagging means simply copying over the sequence shifting subsequent values so that they appear next to preceding values. This permits the estimation of the autocorrelation using the sample correlation. The sample correlations of the three sequences are 0.887, 0.394, and 0.210 for the skip 0, 5, and 10 lag series, respectively. This shows that the assumption that semi-replicates are IID becomes more reasonable as q increases.

8.5 Problems

1. Discuss the importance of improving the computational efficiency of discrete event simulation techniques as advances in computers continue.
2. Is quasi-Monte Carlo a type of variance-reduction technique? Explain briefly.
3. Discuss the extent to which simulation based on Monte Carlo estimation and pseudorandom numbers is currently obsolete.
4. Are common random numbers less valuable for simulations using variable numbers of pseudorandom numbers than for more simple simulations? Explain briefly.
5. All variance reduction techniques generate alternatives to pseudorandom U[0,1] numbers? Explain briefly.
6. Perform Latin hypercube sampling to estimate the expected sum of registration and voting types for system 2 as described in Sect. 8.1.1.

Table 8.8 30 simulation outputs from hypothetical autocorrelated example

#	Simulation outputs	Skip 5	Skip 10	Skip 0 lag	Skip 5 lag	Skip 10 lag
1	0.6998	0.6998	0.6998	1.2821	10.0630	8.1476
2	1.2821			3.2700		
3	3.2700			5.8924		
4	5.8924			7.9123		
5	7.9123			10.0630		
6	10.0630	10.0630		7.8668	8.1476	
7	7.8668			8.0593		
8	8.0593			9.5424		
9	9.5424			8.5472		
10	8.5472			8.1476		
11	8.1476	8.1476	8.1476	6.8276	4.3926	7.1900
12	6.8276			5.6152		
13	5.6152			5.5145		
14	5.5145			5.6381		
15	5.6381			4.3926		
16	4.3926	4.3926		4.9461	7.1900	
17	4.9461			5.5529		
18	5.5529			6.5771		
19	6.5771			6.8962		
20	6.8962			7.1900		
21	7.1900	7.1900	7.1900	7.3817	8.8983	13.5480
22	7.3817			9.2480		
23	9.2480			9.3131		
24	9.3131			9.2644		
25	9.2644			8.8983		
26	8.8983	8.8983		11.0908	13.5480	
27	11.0908			11.7383		
28	11.7383			13.7663		
29	13.7663			12.3582		
30	12.3582			13.5480		
31	13.5480	13.5480	13.5480	11.2260	12.8061	9.9685

7. Perform descriptive sampling to estimate the expected sum of registration and voting types for system 2 as described in Sect. 8.1.1.

8. Perform quasi-Monte Carlo sampling to estimate the expected sum of registration and voting types for system 2 as described in Sect. 8.1.1.

9. Does the sample correlation tend to decrease as q increases, where q is the number of skipped observations from an autocorrelated series? Explain briefly.

10. Consider the sequence $y(i + 1) = (0.9)y(i) + \varepsilon$, where ε is normally distributed with mean zero and standard deviation 1.0. Estimate the smallest number of skipped observations such that the lagged sample correlation is less than 0.25.

Chapter 9
Simulation Software and Visual Basic

The purpose of this chapter is to help the reader learn fundamental skills to program in Visual Basic (VB). Also, the relatively advanced VB code needed to run a discrete event simulation controller is also covered. The transition to the more advanced coding is admittedly abrupt but the code described is self-sufficient and has been used for real scientific computing applications relating to election systems.

There are many programming languages that can be used for discrete event simulation. These include the "high-level languages" associated with commercial software, so-called because they permit the user to rapidly develop code by leveraging built-in templates for queues, processes, and entities. For example, SIMAN is the high-level language associated with the software package ARENA described in Chap. 10

Low-level languages are generic (useful for any type of computing) and include: BASIC, C# (C sharp), C++, Fortran, JAVA, and Visual Basic. Each of these requires the reader to develop event controllers and other constructs, perhaps building on code and architectures in the literature (e.g., see Mesquita and Hernandex 2006, through http://msdn.microsoft.com/ or through one of many excellent message boards accessible by searching the internet for specific VB expressions). These low-level languages offer at least two kinds of benefits:

1. *Low licensing costs*: low-level coding requires little (or no) expense in software licenses. Industrial software licenses for professional simulation software can range from $10,000 to 50,000 per user each year. At the same time, the Visual Basic (VB) language described in this chapter can be used either with an existing license for Microsoft Excel (no additional cost) or for free using the current Microsoft Express Edition Visual Basic compiler. This was the primary reason that my colleagues and I chose to use C++ to write the simulation software used in our voting systems studies described in Chap. 7.
2. *Close-coupling*: when custom-building an application, the opportunity arises to leverage pre-existing optimization or other capabilities such as the API built into many commercial software packages. Exploiting APIs can mean that

T. T. Allen, *Introduction to Discrete Event Simulation and Agent-based Modeling*,
DOI: 10.1007/978-0-85729-139-4_9, © Springer-Verlag London Limited 2011

programs can interact efficiently with the software already in common use in the relevant organization. In general, by building simulation into an excel spreadsheet, the user can exploit the functionalities built into Excel and most easily interface with available data which are often already in Excel spreadsheets. Furthermore, Excel has built-in capabilities for random number generation and specific distributions that reduces development times. The ability to link with existing data and our own simulation–optimization techniques from Chap. 5 motivates our use of Visual Basic code in our on-going hospital modeling studies.

This chapter describes simulation using Visual Basic that focuses on Microsoft Excel-based interfaces. Yet, virtually all of the methods described can run in any Visual Basic environment. Technically, Visual Basic uses an "integrated development environment" (IDE) from Microsoft for its component object model (COM) programming model. COM is an umbrella term referring to many technologies for "inter-process" communication, i.e., a program calling upon another program such as Excel. Visual basic was derived from BASIC and allows comparatively rapid application development of graphical user interface (GUI). In the subjective experience of the author, code development using Visual Basic is generally comparatively pleasant compared with C++ or Fortran, although each language has advantages.

9.1 Getting Started

To open the type of Visual Basic (VB) integrated development environment (IDE) considered here, one must access Microsoft Excel. Here, we focus on Excel 2007 but virtually all methods run on version 2003 or later. (In fact, the author generally uses *Excel 2003* and VB6 since some client organizations have chosen not to upgrade.) Of course, excel is opened either by a desktop short cut or going to *Start > All Programs > Microsoft Office > Microsoft Office Excel 2007*. Once Excel is opened go to the "Office" button (upper left corner of the screen) and select "Excel Options" in the bottom right hand side of the drop down office button menu.

For the 2003 version, Tools > Macro leads to the VB editor and other relevant options. For the 2007 Excel version look within the options "Popular" area be sure that the "*Show Developer tab in the Ribbon*" check box is checked, then click "Ok" to exit the Excel options. Inside the developer ribbon in Excel click the Visual Basic button on the far left-hand side (Note: make sure macro security is enabled). The Visual Basic window will open and look like Fig. 9.1.

Within the upper white space on the left-hand-side below the "Projects—VBAProject" heading, right-click and go to *Insert > Module*. This will turn the major grey area into a white type-able area permitting the writing of VB code. Note that, when saving the document, save it as a "Macro-Enabled" Excel worksheet.

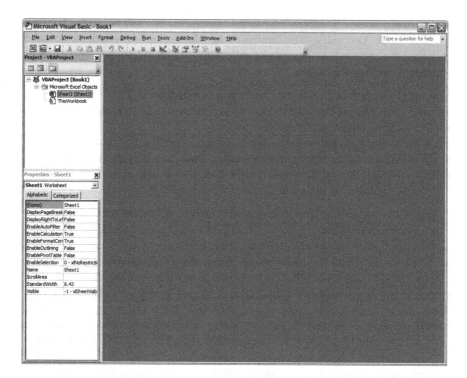

Fig. 9.1 Visual Basic Window

9.1.1 Making a Simple Program

This section describes a simple program that reads in a number from your Excel sheet, multiplies it, and outputs the result back to Excel. Reading in cells to Visual Basic from Excel is simple and illustrates the close-coupling benefit of VB. A code that purely reads in a cell is written as in Code 9.1. Note that line breaks are generated by the "Enter key" and not separators like ";" as in C# or C++. This makes VB more like Fortran. However, there is no necessary space for statements to begin, i.e., they can be staggered visually. Also, capitalization is up to the developer but it is important, e.g., IRead is a different variable than iRead.

Code 9.1 Reading into VB

```
Sub reading_cell()
    Dim iRead As Integer
    iRead = Sheet1.Cells(1, 1)
```

In Code 9.1, "Sub" declares the name, arguments, and code that form the body of a Sub procedure. Once the "Sub" is declared by hitting "Enter" VB automatically writes "End Sub". All of the associated code will be written between the "Sub" and "End Sub". "Dim" refers to the "dimension" or type of the declared variables for internal storage space allocation. Available types appear in a pull-down menu and include:

- Integer—stores whole numbers and negative numbers,
- Long—like integer but with storage space for larger (32 bit) integers,
- Float—stores continuous numbers,
- Double—like float but with storage space for larger (62 bit) numbers, and
- String—stores up to 126 characters from a character set (e.g., ASCII).
- Public—modifier so the variable can be used by all subroutines and functions.
- Private—modifier so the variable has restricted availability.

Generally, the memory requirements for Long and Double data types are not overly burdensome so they are preferred. Also, if variables are undeclared or declared as "Variant," relatively large spaces are needed. For clarity and quality assurance, often "Option Explicit" is used to force the declaration of all variables in the program.

The "iRead" is the name of the variable set by the user, and the Integer means it was set to an Integer number. At the time of definition, its default value is set to zero. iRead, our integer variable, is then set to equal the value in cell A1 in the "Sheet 1" worksheet by the code Sheet1.Cell(1,1). Notice that in the code the column number is not longer done by alphabet but numerically, e.g., that A = 1, B = 2, C = 3.

Next, one can perform a task with the integer value "iRead" that was input. For example, we can multiply the value in cell A1 by 8 then output the result in cell B1. Code 9.1 shows the associated code.

The resulting code can be run by clicking "Macro" under the developer tab in the ribbon. Next, one can select the reading_cell() macro. Also, in the Visual Basic editor, one can simply select "Run" under the "Run" menu, push F5, click the play button (), and/or using the debug menu options such as Ctrl-F8. Selecting run to cursor (Ctrl-F8) permits running the program up to the cursor. Dwelling the mouse over the elements of the code then shows their current values.

TIP: It is generally desirable to use long, descriptive variable names. This makes reading the code easier for others and facilitates search and replace. Also, integers and longs start in many conventions with the letter i.

9.1.2 Other Ways to Interact with Excel

Visual basic code generated through the Excel IDE, although it is associated with your Excel sheet and stored with it, does not interact with Excel unless there is an explicit reason. Explicit ways to interact with Excel sheets extend beyond accessing specific cells. They include:

- Setting a short-cut key,
- Recording a macro, and
- Inserting a command button into the sheet.

Specifically, to create a button in your Excel sheet, one can select (on the developer ribbon) *Insert > Button.* Once the size of the button is created a window will pop up for you to select a macro, which contains VB code. The code will run the macro that was selected every time the button is clicked. For example, a button placed in an Excel sheet might be linked to run the "multiply_cell" subroutine in Code 9.2.

Code 9.2 Reading in A1 and printing

```
Sub multiply_cell()
Dim iRead As Integer
iRead = Sheet1.Cells(1, 1)
Sheet1.Cells(1, 2) = iRead * 8
End Sub
```

Recording macros (which can be a way to implement short-cut keys) offers advantages. For example, it can be a quick way to generate the VB code for operations that the developer already knows well in Excel. As an example, consider how the "multiply-cell" code can effectively be generated by recording a macro.

To record the relevant macro, under the developer ribbon in Excel click "record macro". Next, assign a name such as "record_test" (or any name). Then select cell B2 in Excel and type in =A2*8. Back in the developer ribbon, select the "stop recording" macro.

Note that accessing VB as described will create a module containing the code. That code has a specific location (B1) reference since it was generated when recording the macro. The next command focuses on the active cell and multiplies whatever is in the cell to the immediate left (cell A1 in this case) by 8. The code for this macro as written is show in Code 9.3. Note that single " ' "quotes transform whatever follows in the line into a comment rather than a command.

Code 9.3 Recorded macro code

```
Sub Record_test()
' Record_test Macro
    Range("B1").Select
    ActiveCell.FormulaR1C1 = "=RC[-1]*8"
End Sub
```

9.2 Loops: For and Do–While

It is perhaps true that computers derive the majority of their power from automatic looping, i.e., repetitive tasks. Here, we consider similar types of loops: "for" and "do–while" loops. These are the building blocks for the discrete event simulation code that follows. For loops are generally preferred in cases in which there is a pre-known fixed number of operations; do–while loops which can more easily terminate, are more useful when an operation is to be ended as soon as a condition is satisfied. The code used for illustration here is shown in Code 9.4.

Code 9.4 For loop example

```
Sub Main()
Dim iIndex As Integer, jIndex As Integer
jIndex = 0

For iIndex = 0 To 2
   jIndex = jIndex + iIndex
Next iIndex
End Sub
```

The Code 9.4 is looping "iIndex" from 0 to 2. Anything inside of the "For" and "Next" commands will be performed during each loop. The loop begins with jIndex equal to six and ends when it equals three. A more complicated example of a while loop is shown in Code 9.5.

The code in 9.5 first defined "iIndex" and "jIndex" as integers. It then set the integers to initial values. The while loop then starts and will loop while "jIndex" is less than 4. Each loop the code will subtract one from "jIndex" and add one to "iIndex". When the code is finished it will print "jIndex" in cell A1. Like the "for" loop, the while loop will repeat each task in order during each loop between "While..." and "End While". This executes until the test is false.

Code 9.5 While loop example

```
Sub Main()
Dim iIndex as Integer, jIndex As Integer
iIndex = 0
jIndex = 6
While jIndex > 4
jIndex -= 1
iIndex += 1
End While
sheet1.cells(1,1) = jIndex
End Sub
```

Other variants of looping are shown in Code 9.6 and Code 9.10. Code 9.6 is a general code for the user to fill in the blanks, as it is much like the other loop examples.

Code 9.6 Do loop example

```
Do { While | Until } condition
   [ statements ]
   [ Exit Do ]
   [ statements ]
Loop
```

A "while–wend" is another loop type. An example of this type, in which the loop will repeat 100 times, is shown in Code 9.7.

Code 9.7 While–wend example

```
Dim number As Integer
number = 1
While number <=100
number = number + 1
Wend
```

9.3 Conditional Statements: If–Then–Else and Case

Conditional statements are also core to programming. Here, we consider "If–Then–Else" and "Case" statements. The Code 9.8 illustrates both types of statements. The "If–Then–Else" routes code based on an expression, which is Boolean. If the expression is true, then the first statement is executed, otherwise the else statements (if any) are executed. The "Select Case Variable" construction is similar but its execution depends on the value of the variable.

9.4 Subroutines and Functions

Subroutines and functions are similar constructs that help by organizing the code into specific capabilities. The central difference between subroutines or "subs" and functions is that functions permit information to be passed using the "return value" or function name. Programmers are generally careful about restricting access to variables to avoid mistakes arising from the interaction of functions. As a result, using functions and avoiding public variables (which are not clearly identified with specific functions) is generally preferred.

The code below illustrates two ways of multiplying numbers using subroutines and functions. First, it illustrates how one subroutine can "call" another, interacting through "public" variables. Second, it shows the use of a function and the function return value. Functions can also interact through public variables.

Code 9.8 If–Then and Case Example

```
Sub conditionTest()
Dim iIndex As Long
iIndex = 1
If (iIndex = 1) Then
Sheet1.Cells(1, 1) = "iIndex=1"
Else
Sheet1.Cells(1, 1) = "iIndex=NOT1"
End If
Select Case iIndex
    Case 0
    Sheet1.Cells(1, 2) = "iIndex=0"
    Case 1
    Sheet1.Cells(1, 2) = "iIndex=1"
End Select
End Sub
```

Code 9.9 Illustration of subroutine and function calls

```
Public multiplied As Double
Sub main()
multiplied = 2
Call multiplyByTwo
Sheet1.Cells(1, 1) = multiplied
Sheet1.Cells(2, 1) = multiplyByThree(multiplied)
End Sub
Sub multiplyByTwo()
multiplied = multiplied * 2
End Sub
Function multiplyByThree(number As Double)
multiplyByThree = number * 3
End Function
```

9.5 Visual Basic and Simulation

This section describes code for a discrete event simulation controller. The code can be used to simulate any interarrival distribution, e.g., exponential, lognormal, or other. Also, any service distribution can be used. Further, the code applies to cases with a single queue and an arbitrary number (c) of identical machines or servers. In the queuing terminology of Chap. 6, the code simulates $G/G/c$ systems where "G" stands for general arrivals, general service, and with c machines.

The code is unavoidably complicated. It includes subroutines for manipulating the public variables and functions for changing specific variables. The overall layout is sketched in Fig. 9.2. The simplest components are the initializations and the random numbers functions. Next, the event calls the arrival and department functions and checks for termination conditions. The sections that follow describe each of the sections of the code in detail. All of the code can be pasted into a single module or separate modules.

9.5.1 Pseudorandom Numbers and Initialization

Visual basic contains the built-in random number generator Rnd or Rnd(). With no argument, Rnd returns the next pseudorandom number in the sequence. The quality of the random numbers is not exceedingly high. Pseudorandom numbers from Numerical Recipes, e.g., are clearly of higher quality (Press et al. 2007). As evidence, consider that the numbers derived from Rnd have sometimes crashed our log and other functions by delivering a number so close to zero that the log returned and undefined value. As a result, we developed the following function to generate exponentially distributed pseudorandom numbers. The program checks that the derived Rnd is not equal to zero before taking the logarithm.

Similarly, we developed the following code to generate pseudorandom lognormal deviates. Note the reference to "Application.WorksheetFunction. LogInv" which is based on the built-in Microsoft program library.

By typing "Application.WorksheetFunction." a list appears of available functions. We have experienced occasional issues with not being able to run code using these functions. Yet, saving and restarting the program has generally fixed related problems. The following code generates approximate nonhomogeneous arrivals. These numbers are approximate because, as noted in unpublished communications by David Kelton, the expected arrival rate is not exactly the target $\lambda(t)$. Kelton has proposed a so-called "thinning" method based on hypoexponential random

Fig. 9.2 Overview of the discrete event simulation code

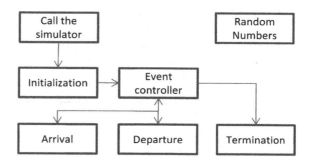

variables (sums of exponentials) with arrival rates that are provably consistent with nonhomogeneous Poisson processes. This more exact generation method is commented in the code.

The code example focuses on assumptions appropriate for some formulations of Election Day simulations. The specific approach hard-codes a doubling of the arrival rate in specific rush periods. Note that the rush periods are assumed to be early in the morning and after work.

Completing the preparation for writing the core program requires initialization. The declaration "Option Explicit" forces the declaration of all variables in the file. The declaration of public variables is important because they are involved in arrivals, departures, and checks for the completion of the simulated period. Almost all of these public variables relate to quantities being tracked related to the evaluation metrics such as the waiting time and the number in queue.

Code 9.10 This generates exponential pseudorandom numbers in a robust way

```
Function ExpoVariate(Rate As Double)
'The challenge is to avoid any possible crash if Rnd gives 0.000.
  Dim UniVariate As Double
  UniVariate = 0
  Do While UniVariate = 0
     UniVariate = Rnd
  Loop
     ExpoVariate = -Log(UniVariate) / Rate
End Function
```

Code 9.11 Robust lognormal random number generator

```
'The following generates lognormal pseudorandom numbers.
Function geneLN(lNMu As Double, lNSigma As Double)
'The challenge is to avoid any crash if Rnd gives 0.000.
   Dim UniVariate As Double, N As Double
      UniVariate = 0
      Do While UniVariate = 0
      UniVariate = Rnd
   Loop
N=Application.WorksheetFunction.LogInv(UniVariate,lNMu,lNSigma)
geneLN=N
End Function
```

Code 9.12 Generates exponential interarrivals for nonhomogeneous Poisson processes

```
Function NonStationaryExpo(Rate As Double)
'Assume two peaks, i.e., 6:30 - 8:30, and 17:30 - 19:30.
'The arrival rates in those peaks are twice
'of those in other periods, i.e., 8:30-17:30
Dim MaxRate As Double
MaxRate = 26 / 17 * Rate
'Rate here is the arrival rates in the stationary case
    If clocktime <= 120 Or clocktime > 660 Then
        NonStationaryExpo = ExpoVariate(MaxRate)
    End If
        If clocktime > 120 And clocktime <= 660 Then
NonStationaryExpo = ExpoVariate(MaxRate/2)
        End If
'The following algorithm uses the "thinning" method
'    If clocktime <= 120 Or clocktime > 660 Then
'        NonStationaryExpo = ExpoVariate(MaxRate)
'    End If
'    If clocktime > 120 And clocktime <= 660 Then
'        Do
' NonStationaryExpo = NonStationaryExpo + ExpoVariate(MaxRate)
'        Loop While Rnd > 1 / 2
'    End If
End Function
```

Also, "Option Base 1" initializes arrays with the value 1. Further, the "Infinity" variable will store a large value used in specific initializations. The parameters "lambda" and "mu" are also used throughout. Note also the extensive commenting to help others use the code without detailed explanations.

9.5.2 The Event Controller

This section describes the functions associated with the event controller. As noted in Chap. 4, the heart of the event controller lies in enumerating the possible next occurrences. Then, it identifies the time of the next occurrence and moves the clock forward. For the single queue simulation considered here, there are only three possibilities for the next event:

1. An arrival,
2. A departure (from any server), and
3. Termination of the program.

The first function determines the next event. It searches the next arrival, the jobs being performed (if any), and checks for termination. Note the initialization using the infinite global variable to facilitate the selection of the shortest time derived from the "for" loop. Because the number of servers is generally small, it is appropriate to use an Integer for the index in the loop.

Code 9.13 Initial declarations for a discrete event-simulation controller

```
Option Explicit
Option Base 1
Public NextArrival As Double
Public NextDeparture() As Double
Public NumInSystem As Integer
Public NumInQueue As Integer
Public ServerBusy() As Boolean
Public lambda As Double, mu As Double
Public Server As Integer
Public Infinite As Double, StartToWait As Double
Public cumWaitTime As Double
Public Numserver As Integer, closetime As Double
Public clocktime As Double
Public MaximumVoter As Integer
Public Queue_ArrivingTime() As Double
Public CurrentVoterWaitTime As Double
Public CurrentMaxWaitTime As Double
Public cumCurrVoterWaitTime As Double
Public testcounter As Integer
```

Code 9.14 This function finds the time of the next event

```
Function findclocktime() As Double
Dim MinDeparture As Double
Dim i As Integer
MinDeparture = Infinite
   For i = 1 To Numserver
      If MinDeparture > NextDeparture(i) Then
         MinDeparture = NextDeparture(i)
         Server = i
      End If
   Next i
   If NextArrival > MinDeparture Then
      findclocktime = MinDeparture
   Else
      findclocktime = NextArrival
   End If
End Function
```

The main body of the event controller function is long and unavoidably complicated. As a result, it is divided into three parts in this text. In general, subroutines ("subs") and functions can be deployed among different modules with no effect. This permits the organization of large codes. The three portions of the event controller function described here are different parts of the same function and thus must be in the same module.

The controller code begins with more initializations. The variables that track time, waits and line lengths are set to zero. The code is designed to be flexible such that tracking can be begin with zero-length queues or after the queues have been allowed to build. Waiting would help for modeling steady state periods as described in Chap. 5. A "burn-in" period is often used in these cases to address the initial conditions bias. In the context of election systems simulation, starting the clock at the opening of the polls might make sense since the election officials might not be held accountable for lines formed while the polls were closed.

The next section focuses on the initialization of counters for cases in which variables were set externally such that a line has already formed because of initializations. First, if there is a line the next departure time is calculated. All servers are initialized as busy. Alternatively, if the number in the system is less than the number of servers, the next departure issue is unknowable so it is set to a large number (Infinite).

In the last section of the event controller, initialization is complete. The "Do While" essentially controls the simulation. The next event determines the event type. The event type and the case statement kick off the execution of the event type. After executing the arrival or departure event using the functions described in the next section, records are updated. These permit estimation of the average waiting time of all voters being served.

The record keeping also permits computation of the average long-term waiting time. The program refers to this long-term average as the "Max waiting time" and changing the commenting allows this quantity to be calculated instead of the average waiting time for all voters.

9.5.3 Arrival and Departure Events

The subroutines that describe arrival and departure are next described. First, when an arrival occurs, the waiting time counter must be initialized. If waiting were needed, then the queue length would also need to be lengthened. Alternatively, if there is no line, then the first available machine needs to be loaded with the job (or voter). If loaded, we can immediately calculate the departure time of the loaded job. The global storage variables are adjusted.

Code 9.15 The start of the event controller code

```
Function AverageWaitTime(servicerate As Double, NumEarlyVoter As
Integer, TotalVoters As Integer)
  Dim i As Integer, j As Integer, Rep As Integer
  Dim EventType As Integer, variance As Double
  Dim stdev As Double
  Dim TotalServed As Integer, WaitTime() As Double
  Dim TotalWaitTime As Double
  Dim Voter As Integer, cumqueue As Integer
  Dim averagequeue As Double
  Dim TotalMaxWaitTime As Double, MaxWaitTime() As Double
  Infinite = 60 * 60
  Rep = 100
  ReDim WaitTime(Rep)
  'The average wait time of one replication
  ReDim MaxWaitTime(Rep)
  MaximumVoter = TotalVoters
    For j = 1 To Rep
      closetime = 13 * 60
      clocktime = 0
      mu = servicerate
      NextArrival = NonStationaryExpo(lambda) 'ExpoVariate(lambda)
      NumInSystem = NumEarlyVoter
      Voter = NumEarlyVoter
      cumWaitTime = 0
      StartToWait = 0
      TotalServed = 0
      ReDim NextDeparture(Numserver)
      ReDim ServerBusy(Numserver)
  'Depends on whether the queue begins with empty or not.
      If NumInSystem = 0 Then
        For i = 1 To Numserver
          NextDeparture(i) = Infinite
        Next i
        NumInQueue = 0
      End If
```

Code 9.16 The middle of the event controller code

```
If NumInSystem >= Numserver Then
For i = 1 To Numserver
NextDeparture(i) = clocktime + (1 + mu) * geneLN(1.7042, 0.4406)
'ExpoVariate(mu)
    ServerBusy(i) = True
Next i
    NumInQueue = NumInSystem - Numserver
End If
If NumInSystem < Numserver Then
For i = 1 To NumInSystem
NextDeparture(i) = clocktime + (1 + mu) * geneLN(1.7042, 0.4406)
'ExpoVariate(mu)
    ServerBusy(i) = True
Next i
For i = NumInSystem + 1 To Numserver
    NextDeparture(i) = Infinite
Next i
    NumInQueue = 0
End If
If NumInQueue > 0 Then
    ReDim Queue_ArrivingTime(NumInQueue)
Else
    ReDim Queue_ArrivingTime(1)
End If
    clocktime = findclocktime
```

Second, the arrival event subroutine is described. If there is no queue, the departure has limited implications. The server is not used and the time of next job completion is unknown. Alternatively, if there is a queue the departure means there will be an immediate replacement on the machine. This requires updating the waiting times and the times of the next departures. The stack of processing jobs needs to be shifted to account for the removal of the completed job.

Code 9.17 The end of the event controller code

```
Do While clocktime < closetime Or NumInSystem > 0  'And Voter <= TotalVoters
        If clocktime = Infinite Then
           Exit Do
        End If
        If clocktime = NextArrival Then
           EventType = 1
        Else
           EventType = 2
        End If
        Select Case EventType
           Case 1
              Voter = Voter + 1
              'Call arrival
              'If clocktime < closetime Then Voter = Voter + 1
              If Voter <= MaximumVoter Then
                 Call Arrival(Voter)
                 'Debug.Print Voter
              End If
              'Range("D2").Cells(Voter).Value = cumWaitTime
           Case 2
              Call Departure(Server)
              TotalServed = TotalServed + 1
        End Select
        If TotalServed > 0 Then
'WaitTime(j) = cumWaitTime / clocktime  'This is averge number in queue
           WaitTime(j) = cumWaitTime / TotalServed
        End If
        clocktime = findclocktime
     Loop
     TotalWaitTime = TotalWaitTime + WaitTime(j)
     MaxWaitTime(j) = CurrentMaxWaitTime
     'the sum of Max wait time
     TotalMaxWaitTime = TotalMaxWaitTime + MaxWaitTime(j)
     CurrentMaxWaitTime = 0
  Next j
  'Expected wait time
     AverageWaitTime = TotalWaitTime / Rep
     'Calculate standard deviation
     'For j = 1 To Rep
        'variance of the expected wait time
        'variance = variance + (WaitTime(j) - AverageWaitTime) ^ 2
        'variance of the max wait time
        'variance = variance + (MaxWaitTime(j) - AverageWaitTime) ^ 2
        'Range("a2").Cells(j).Value = WaitTime(j)
     'Next j
     'stdev = (variance / (Rep - 1)) ^ 0.5
     'Debug.Print AverageWaitTime; 'stdev; stdev / AverageWaitTime
  End Function
```

Code 9.18 This executes an arrival event

```
Sub Arrival(ArrivedVoters As Integer)
Dim ToDoServer As Integer, i As Integer
Dim DepartureIncre As Double
Dim ArrivalIncre As Double
  'clocktime = NextArrival
  NumInSystem = NumInSystem + 1
  If NumInSystem > Numserver Then
    'compute cumulative wait times
cumWaitTime = cumWaitTime + (clocktime-StartToWait)*NumInQueue
  NumInQueue = NumInQueue + 1
  StartToWait = clocktime
  'record current voters in queue with their arrivingtime
  ReDim Preserve Queue_ArrivingTime(NumInQueue)
  Queue_ArrivingTime(NumInQueue) = clocktime
Else
  'to find an idle server
  For i = 1 To Numserver
    If ServerBusy(i) = False Then
      ToDoServer = i
      Exit For
    End If
  Next i
  ServerBusy(ToDoServer) = True
NextDepature(ToDoServer)=clocktime+(1+mu)*geneLN(1.7042,0.4406)
'ExpoVariate(mu)
  End If
  If ArrivedVoters >= MaximumVoter Then
    NextArrival = Infinite
    Exit Sub
  Else
    NextArrival = clocktime + NonStationaryExpo(lambda)
  End If

  If NextArrival > closetime Then
    NextArrival = Infinite
    Exit Sub
  End If
End Sub
```

Code 9.19 This executes a departure event

```
Sub Departure(Server As Integer)
Dim DepartureIncre As Double
Dim k As Integer
    'testcounter = testcounter + 1
    NumInSystem = NumInSystem - 1
    If NumInQueue = 0 Then
        ServerBusy(Server) = False
        NextDeparture(Server) = Infinite
    Else
        'This computes cumulative wait time.
    cumWaitTime = cumWaitTime + (clocktime - StartToWait) * NumInQueue
        ServerBusy(Server) = True
        NumInQueue = NumInQueue - 1
        StartToWait = clocktime
        'This computes the specific voter's waiting time.
        CurrentVoterWaitTime = clocktime - Queue_ArrivingTime(1)
    cumCurrVoterWaitTime = cumCurrVoterWaitTime + CurrentVoterWaitTime
        'This helps identify the maximum waiting time so far.
        If CurrentMaxWaitTime < CurrentVoterWaitTime Then
            CurrentMaxWaitTime = CurrentVoterWaitTime
        End If
        If NumInQueue > 0 Then
            For k = 1 To NumInQueue
                Queue_ArrivingTime(k) = Queue_ArrivingTime(k + 1)
            Next k
            ReDim Preserve Queue_ArrivingTime(NumInQueue)
        End If
    NextDeparture(Server) = clocktime+(1  +  mu)*geneLN(1.7042,  0.4406)
    End If
    End Sub
```

9.5.4 Calling the Average Waiting Time Function

Finally, there is a simple subroutine that calls the function described in previous sections. The default parameters are selected in the code below to represent a 13-h election day. Also, the arrival rate and the number of machines is intended to describe a medium-sized voting precinct in terms of the expected number of entities (registered voters) arriving. In the example, the number of voters arriving is assumed to be 433 over 13 h. The service rate is 0.384 entities (voters) per minute. The function then writes the average waiting time to the first cell in Sheet1 in the Excel workbook.

Code 9.20 Example code for calling the discrete event controller

```
'For the use of debugging
Sub testWaitTime()
    Dim mu As Double
    lambda = 443 / 13 / 60
    mu = 0.384
    Numserver = 8


    Call AverageWaitTime(mu, 36, 443)
    Sheet1.Cells(1,1) = AverageWaitTime

End Sub
```

9.6 Problems

1. List three advantages of using low-level languages and custom simulation software instead of commercial software.
2. List one advantage of VB compared with an alternative language.
3. Is it possible to us VB code independent of Microsoft Excel using a Microsoft controller and paying little or no licensing fee? Explain briefly.
4. What dimension types are the most economical choices possible for the following numbers: 0.2123456233, 0.231, 32, fun, 3212343234.
5. What dimension types are the most economical choices possible for the following numbers: 0.2123, 0.231, 321232112, {fun, time}, 33234.
6. Record a macro that multiplies a number by 27.
7. Record a macro that divides a number by 27.
8. Write code that sums $1 + 2 + 3 + 4 + \cdots + 100$.
9. Write code that sums $1 + \frac{1}{2} + 1/3 + \frac{1}{4} + \cdots + 1/100$.
10. Write code to calculate $1 + 2^2 + 3^3 + \cdots$ (the largest number in the series <1000).
11. Write code to calculate $1 + 2^2 + 3^3 + \cdots$ (the largest number in the series <1000).
12. Write VB code to create a "smiley face" made out of yellow Excel cells.
13. Write VB code to color five cells yellow in Excel.
14. Write VB code to create the sequence $\ln(i)$ for $i = 1,\ldots,10$ or "i = 5" for $i = 5$.
15. Write VB code to create the sequence $\exp(i)$ for $i = 1,\ldots,10$ or "i = 5" for $i = 5$.
16. Repeat question 14 using a While–Wend loop.
17. Repeat question 15 using a While–Wend loop.

18. Assume that X is exponentially distributed with mean 10. Use VB code to estimate $E[X^2]$.
19. Assume that X is exponentially distributed with mean 20. Use VB code to estimate $E[X^2]$.
20. Assume that X is lognormal with parameters 2 and 3. Use VB code to estimate $E[X^2]$.
21. Assume that X is lognormal with parameters 2 and 3. Use VB code to estimate $E[X^2]$.

Assume that customers arrive over a 30 min period according to a Poisson process with mean interarrival time of 2 min. Also, assume that service is lognormal with parameters 1 and 1.5. Use VB code to estimate the average waiting time.

Chapter 10
Introduction to ARENA Software

This chapter introduces a software package used widely for instruction and real-world decision-support often referred to as ARENA software. ARENA is effectively a suite of software which includes the ARENA simulator, the Input Analyzer (for input analysis), and the Process Analyzer (PAN) (for output analysis). All of these are produced by Rockwell International which is also a major manufacturer. This situation permits the ARENA team to constantly apply their software to real-world problems. ARENA software and the related suite do not have the full range of graphics and visualization capabilities of software such as AutoMod, MODSIM, PROMODEL, SIMIO, and many others. Also, the Input Analyzer and the PAN lack statistical analysis features in more powerful software such as SAS. Yet, ARENA is comparatively easy-to-learn, of moderate cost (around $20K per a professional license), and can be successfully used for even large projects.

The purpose of this chapter is to help the reader create and run a simple ARENA model. Also, the chapter orients the reader to the other software in the ARENA suite. Section 10.1 describes in detail how to model the election system example from Chap. 4 using ARENA. Section 10.2 applies the Input Analyzer to a related data set and Sect. 10.3 explores the application of the PAN. For those professionals who prefer accessing their models through a programming environment, ARENA also provides a programmatic interface through a language called SIMAN that we will describe only briefly in Sect. 10.4. Section 10.6 concludes with an example that summarizes the key methods and concepts from the chapter.

This chapter covers only the three most basic ARENA "modules" or "blocks" shown in Table 10.1—Create, Process, and Dispose. Together, these can create the entities governed by a random number stream and the event controller. The Process module can be thought of as a bank of machines with a queue. When an entity arrives and a resource is available, the entity seizes the resource, delays the resource from receiving other entities in queue (if any), and then (after service completes) releases the resource. Chapter 11 describes more advanced modules for scheduling, material handling visualization, and general animation.

T. T. Allen, *Introduction to Discrete Event Simulation and Agent-based Modeling,*
DOI: 10.1007/978-0-85729-139-4_10, © Springer-Verlag London Limited 2011

Table 10.1 ARENA modules used in this chapter

Create	This module is intended as the forming point for entities in a simulation model. Entities are created in one of two ways: (1) a schedule (2) based on a time or expression between arrivals
Process	This module is the main processing method in the simulation. Options for seizing and releasing resource constraints are available. Normally, an entity seizes, delays, and then later releases the process or machine. The process time is allocated to the entity and may be considered to be value added, nonvalue added, transfer, wait or other for lean production related accounting
Dispose	This module is intended as the ending point for entities in a simulation model. Entity statistics may be recorded before the entity is disposed

10.1 Getting Started: Voting Example

The voting system has been described in this book in Chaps. 1, 3 and 8. The relevant workflow is detailed in Fig. 3.3, which already is effectively an ARENA-type model layout. We are going to create this model using the above-mentioned blocks. Starting with a Windows® computer that has ARENA installed on it, double-click the ARENA icon on the desktop to start the program. Alternatively, use the start menu to go to *Programs > Rockwell Software > Arena* (with the appropriate version number) and click on the ARENA icon. A result similar to Fig. 10.1 will appear.

Next, study the areas of the page you have opened. Versions differ somewhat. Generally, the words "Basic Process" will be displayed on the left side of the screen below which will appear the Project Bar. Inside the project bar are the blocks known as modules. Below these modules is a Reports bar. This facilitates the viewing of reports derived from a previously run simulation. Below the reports bar is the navigation bar which can be used to navigate around your model window (see description below). If you wish to go back to the basic process bar you click on it. To the right of the Project Bar the model window is displayed in which the simulations are created. The model window dominates visually.

To create the voting system model, drag and drop a Create module from the basic processes to the model window. Double click the "Create 1" module now sitting in the model window, an image similar to Fig. 10.2 should appear.

Based on the data in Fig. 3.3, we need to change the value and units entries in this box to reflect an average interarrival time of 0.2 min. The resulting user form should look like Fig. 10.3.

After making these changes, click OK. Next drag and drop a Process module to the model window. Note that the queuing will be preformed along the path ARENA created between the Create and Process modules. Double click your

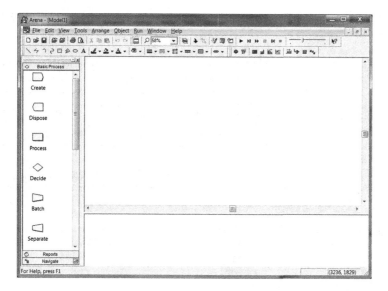

Fig. 10.1   ARENA blank document

Fig. 10.2   Create 1 module open

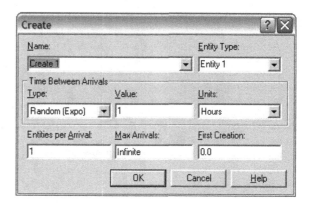

Process module to enter information into the associated user form. The result should look like Fig. 10.4.

Change this box as follows. Note that when you change the action from "Delay" to "Seize Delay Release" you need to click "Add" next to the "Resources:" box. The default setting that appears is acceptable in this case, therefore click ok. Almost always, we select "Seize Delay Release" and to add a resource. Without making both of these choices, the model would not run (no resource) and if it did, that resource would never be released resulting in a generally accumulating queue (Fig. 10.5).

Click "OK" to exit the Process module. Next, we will create another Process module, repeating the steps just carried out for the arrival and registration modules. Here, "creating" implies dragging and dropping a module from the

Fig. 10.3 Create module
updated

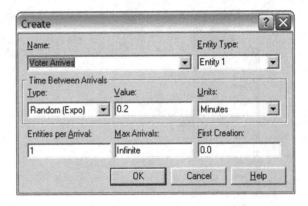

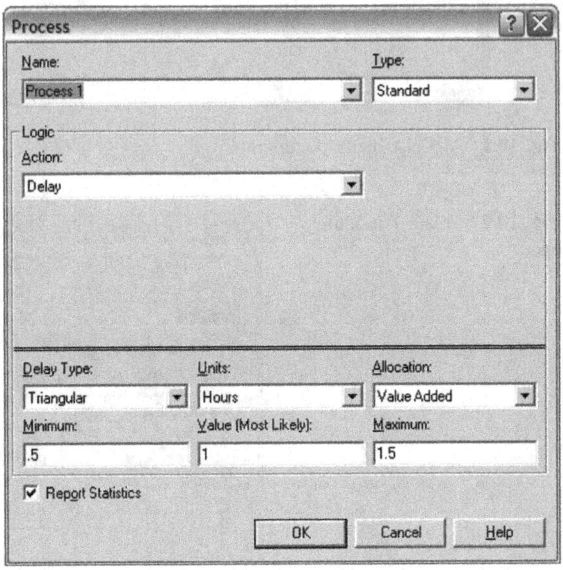

Fig. 10.4 Process module
opened

project bar into the model window. This was done earlier, we created the create (arrival) and process (registration) modules. Select the new module and change the user form it so it looks like Fig. 10.6. Again, "Seize Delay Release" is used and a resource is created by adding it.

Following the completion of the Voting Machine process create a "Dispose" module and leave the default settings. Your model should look similar to Fig. 10.7.

If your figure has connections going to the wrong locations, or is out of order, you can simply click the connections and press delete on your keyboard. Then, pressing the ⬚ button you can create new paths by clicking on the end of the module where one would wish to start followed by clicking on the end of the module where one would like to finish. Lastly in the model creation go to

Fig. 10.5 Process module
updated

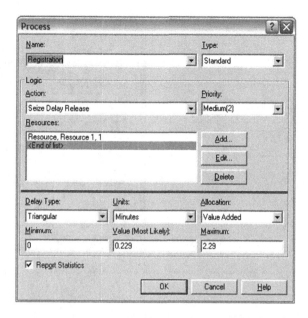

Fig. 10.6 Process module 2
updated

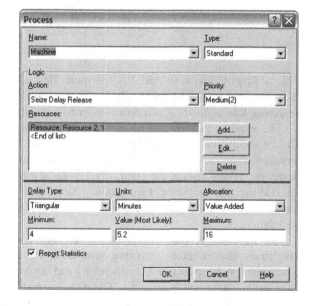

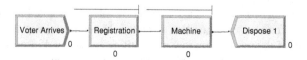

Fig. 10.7 ARENA voting system layout

Fig. 10.8 Run setup window

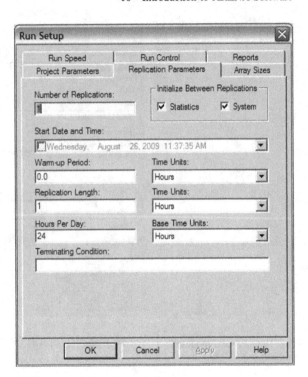

File > Save As and save ARENA file (note files save using the .doe extension). To run the model one must adjust the run settings. Generally, this requires selecting *Run > Setup*. A window similar to Fig. 10.8 appears.

The window in Fig. 10.7 allows the selection of the number of replications for the simulation including the time length of the simulation, etc. In the voting systems simulation, one would want to specify the run time as the number of hours the voting booths are open, say 1 h for a trial run (or 13 h plus overtime in realistic cases). Once the running parameters have been set in the run setup window, press the ► button on the main ARENA window to run the simulation.

We have set up the simulation to run where the capacity for resource 1 (registrar) and resource 2 (voting machines) are 5 and 35, respectively. The simulation is set up to run 25 replications and our voting day is set at just 1 h so that results come nearly instantly. Once the program is run, results like those in Fig. 10.9 will appear on your screen after the viewing reports prompt is answered with a yes response.

As shown in Fig. 10.9 the average number of voters who voted in the 1-h voting window for the current configuration over the 25 replications is 215. Under the "Preview" tab of the reports window, the entities, queues and resource figure data are displayed. Assuming a reasonable number of replications (such as 25) is used, the program outputs will include estimated means and confidence interval half

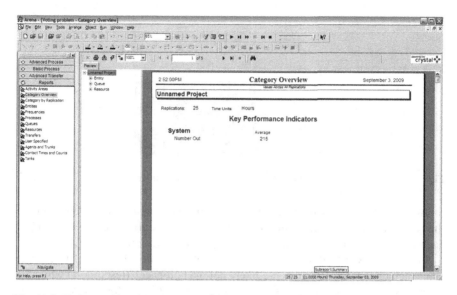

Fig. 10.9 Voting problem initial report screen

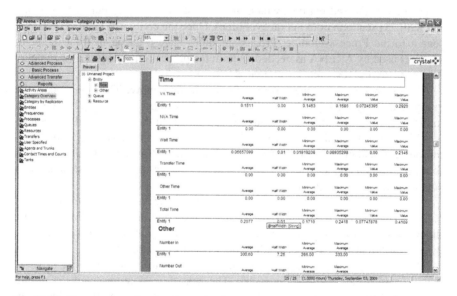

Fig. 10.10 Report output showing the entity times

widths as described in Chap. 2. For example, Fig. 10.10 shows the times for the
entities in the following areas: value added (VA) time, nonvalue added (NVA)
time, waiting time, transfer time, other time, and total time though out the system.

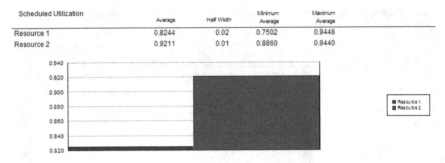

Scheduled Utilization	Average	Half Width	Minimum Average	Maximum Average
Resource 1	0.8244	0.02	0.7502	0.9448
Resource 2	0.9211	0.01	0.8860	0.9440

Fig. 10.11 Report output showing estimated resource utilization with half width

Additional results of interest are displayed in the comparison bar graph inside the resource section of the report. This chart shows the utilization (see Chap. 6) of each resource compared visually. An example is shown in Fig. 10.11.

Consider how the output analysis data might be applied. From Fig. 10.11, one can conclude that the registrars are likely not the bottleneck (resource 1) while the machines (resource 2) are. This finding suggests opportunities for system redesigns as described in Chap. 7.

Having described the basics for creating a working ARENA model, we briefly review the underlying logic. We used a "Create" that we called "Voters Arrive" to generate simulated entities, which in this case were voters. For this, we used EXPO(0.2) pseudorandom interarrival times. Once the entity was created, ARENA sent it to the Process module. Process modules have a queue in front of them, which is used by ARENA as needed. Within the process one chooses a seize delay release using the resource 1 we created.

The "seize" reserves the resource for the loaded entity. The delay takes the entity and delays it based on the input, which was a TRIA(0, 0.229, 2.29) in minutes. During this time the resource was utilized. In the example, this resource was chosen to be used only once. If one wished instead to have, e.g., four registration representatives then the value of "Resource 1 Capacity" would be set at four. In common applications, we would not add four resources but instead apply a single resource with "Quantity" or "Capacity" equal to four.

In a more complex problem, it is generally advantageous to name the resource, e.g., the registration representative, for accounting and debugging. For example, there might be "Decide" module (see ARENA help and Chap. 11) that could route parts to different resources depending on a condition being met.

The "release" frees the resource and the entity to go to the next process, where again it can queue if necessary. The next process represented the voting machines. This module represents the person actually voting. Again a seize delay release is used, this time we input a TRIA(4, 5.2, 16) distribution in minutes. And the resource 2 represents a single voting machine that was inputted. Finally, the last module is a dispose which removes said entity from the simulation system.

10.2   ARENA Input Analyzer

ARENA also has available in its tool box an Input Analyzer as described in Sect. 2.3. This program takes as input the data typically collected in a text file. It then generates the best fit curve and expression to enter into the associated Create, Process, or other ARENA block. For example, in the voting problem in Sect. 10.2, two triangular distributions, and a single exponential distribution were derived from fitting data.

To open the Input Analyzer, go to *Start > Programs > Rockwell Software > ARENA > Input Analyzer*. Once the program is opened go to *File > New* or click the ☐ button. The data must be in .txt format with space, tab, or line delimiting. (Typically the needed inputs have been prepared by copying and pasting data from a word processing or spreadsheet program directly into a notepad .txt file.)

To open and analyze the data, go to *File > Data File > Use Existing...* select the file type as .txt then locate and double click your saved .txt file with the data. Opening a .txt with the following set of data (90, 95, 90, 90, 98, 92, 93, 94, 91, 88, 89, 91, 92, 87, 90) and selecting *Fit > Fit All* should yield a result similar or identical to Fig. 10.12 (depending on versions).

The Input Analyzer fits all of the following distribution types: beta, empirical, erlang, exponential, gamma, Johnson, lognorm, normal, poisson, triangular, uniform, weibull. Fits are based on a relative frequency histogram and the sum of squared error criterion as described in Chap. 3. Within the Fit tab you can choose any one of these options or the fit all and let the analyzer choose for you. Generally, the simplest model with a competitive sum of squares error is preferred.

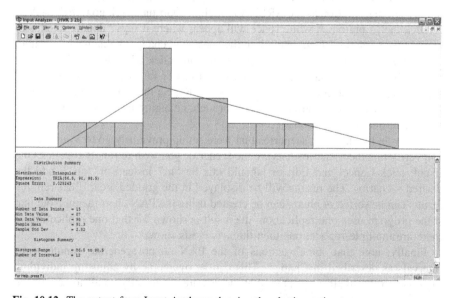

Fig. 10.12   The output from Input Analyzer showing the plotting estimates

As an example output, consider the distribution summary, data summary, and histogram summary shown in Fig. 10.12.

10.3 ARENA Process Analyzer

The PAN helps users in evaluating alternatives to their ARENA base model, e.g., output analysis of how would results change if more voting machine resources were added? The PAN takes as input ARENA files created by the execution of different model scenarios. Its outputs display results so as to facilitate statistical comparison of the models and aid in decision-making.

The ARENA help menu defines the following key terms as follows:

Controls inputs that are considered to affect the operation of the model in a manner that can be monitored/viewed in the output of the model.
Responses outputs that represent measures of how the model performed during the run.
Scenario a collection of controls and responses as applied to a given simulation model.

Therefore, controls are generally either resource capacities or variables (not yet described here). The voting problem used resource 1 as registrars and resource 2 as voting machines; in the example given, each parameter was set to one. To start the PAN, go to the start menu then *Programs > Rockwell Software > ARENA > Process Analyzer.*

The PAN works with .p files. To create a .p file you must run your ARENA model and select the *check model* option in the run menu, ARENA will create a .p file in the same location as the ARENA model. To start the PAN, go to *File > New* or click the ☐ button. A white space will appear where it says "double-click here to add new scenario." Double click in the space and change the names as desired.

Next, browse for the .p file you wish to use and click OK. Then, right click on the newly created scenario and select *Insert Control* to set up a control for this scenario. Change the control settings as desired. Repeat the steps including double-clicking to add new scenarios and step through, changing the control settings for each additional desired scenario.

To re-run the different scenarios that were set up previously go to *Run > Go...* and verify that the scenarios lists are the scenarios you wish to run. If so, click OK; if not, click Cancel and Edit or [should this be "to" instead of "or"?] add the desired scenarios. The results will be displayed in the gridded area. If desired, the results can be sorted or charts can be created using the PAN chart menu. Note that if the program uses one replication, the result is shown for that one replication; if there are multiple replications then the average is shown.

Finally, note that some versions of the PAN do not generate simultaneous intervals as described in Chap. 5. This can be overcome by adjusting individual intervals so that their associated alpha levels are sufficiently small to satisfy the Bonferroni or other rigorous inequality as described in Chap. 5.

10.4 Processes, Resources, Queues, and Termination

In most cases there is more than one way to perform tasks in ARENA. For example, one can use the project bar to create modules in the model as described previously which involved dragging and dropping. Once created the user double clicks the module to make changes.

Another method for revision requires clicking on the Process icon in the Project Bar to display a list of all the process modules as shown in the circled area of Fig. 10.13. Each column in this area is an option that would be found inside the module user form. Viewing the table and making changes there is often more convenient when one has a large simulation model and many changes are needed.

Similarly if you wish to change entity, queuing, resource, variable or schedule settings, one would select the appropriate box in the project bar and change the settings in the bottom area circled in Fig. 10.10. For example, to change the queuing style select the queuing box as shown in Fig. 10.14 for the voting example.

One aspect of queuing that can be changed is the type. Types include: first-in-first-out (FIFO), last-in-last-out (LILO), lowest attribute value, or highest attribute value. These choices determine which entities are allowed to leave the queue first, if and when a resource becomes available.

Additionally, if you wish to change the ARENA animated figure for each entity you can select the entity box in the project bar. In the column "Entity Picture" you can use the drop down menu to select the desired picture. Additional details about animation options are described in Chap. 11.

Finally, when running a simulation there are at least two ways to end the simulation. As previously mentioned one can specify a length in time.

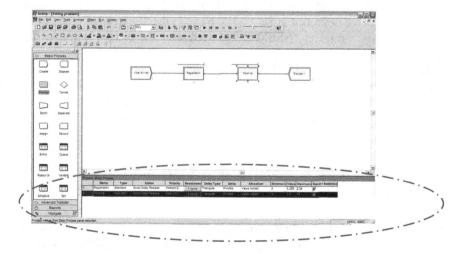

Fig. 10.13 The ARENA environment showing the bottom bar

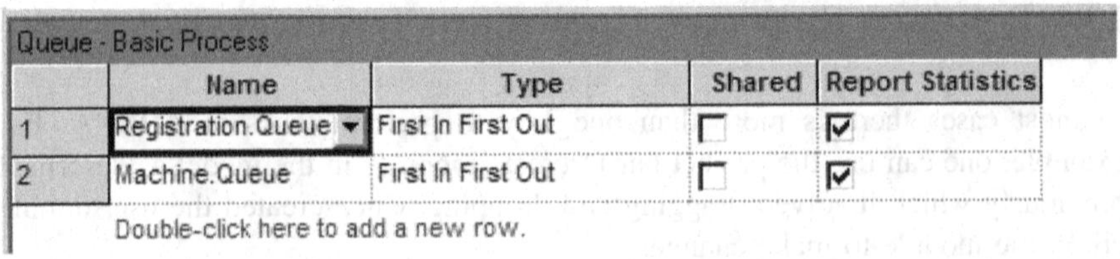

Queue - Basic Process				
	Name	Type	Shared	Report Statistics
1	Registration.Queue ▼	First In First Out	☐	☑
2	Machine.Queue	First In First Out	☐	☑
	Double-click here to add a new row.			

Fig. 10.14 Dialog box permitting the entry of queue parameters

Alternatively, one can set a specific condition, e.g., if a specific queue length goes to zero. Terminating conditions can be entered using the *Run* > *Setup* user form.

Experimenting with ARENA is a good way to learn it. Small models rarely crash the program. As previously mentioned there are usually numerous ways to perform functions. Remember that the ARENA built-in help menu offers a valuable resource.

10.5 Nonhomogeneous and Batch Arrivals

It is common that arrivals are more likely during certain periods than others, i.e., arrivals are "nonhomogeneous" as described in Sect. 4.1. Also, it is somewhat common that when entities arrive, they appear in groups or "batch" arrivals. For example, arrivals at a restaurant are more likely during lunch time than during the morning. Some of the customers will arrive together with friends.

In ARENA, nonhomogeneous and/or batch arrivals can be accomplished through the selection of options in the "Create" block. The process begins with setting up a schedule with the average number of entities arriving in each period. Schedules can be entered under the "Basic Processes" window. Additional details about schedules are described in Sect. 11.3. Next, in the "Expression" field in the "Create" dialog, we select "NSEXPO" and reference the created schedule. The selection is shown in Fig. 10.15.

In addition, the "Create" block permits the number of entities per arrival to be a random variable. These batch arrival options are reasonably accomplished using the (discrete) Poisson distribution. This can be entered using the "POIS(mean)" construction, where mean is the average group size. An example is also shown in Fig. 10.15.

10.6 Summary Example

This example involves the Input Analyzer and the three elementary blocks in ARENA: Create, Process, and Dispose. It also involves a conditional "Decide" block described in more detail in the next chapter.

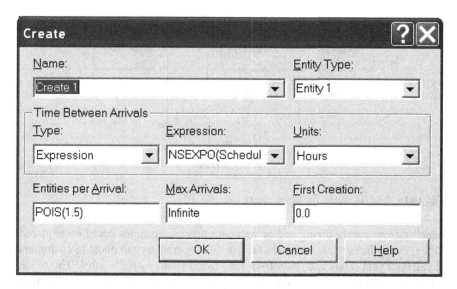

Fig. 10.15 Create block showing nonhomogeneous and batch arrivals

Assume that parts arrive at a single machine system according to an exponential interarrival distribution with mean 20 min; the first part arrives at time 0. Upon arrival, the parts are processed at a machine. The parts are inspected and about 24% are sent back to the same machine to be reprocessed (same processing time). Also, assume that 20 processing times are collected and an exponential distribution is fitted in the Input Analyzer and the p value is 0.03. Most processing times are around 16 min and all are between 11 and 18 min. Questions include:

1. What service time distribution is a reasonable and defendable choice?
2. What is the average number of parts in the machine queue?
3. What is the average cycle time (time from a part's entry into the system until its exit after however many passes through the machine system are required)?
4. What are some possible benefits from having the model?

Answers follow. (1) The exponential distribution is not defensible because the KS-test ruled it out (see Chap. 3). The p value is <0.05 so that the lack-of-fit is significant. Without more information, the assumption TRIA(11, 16, 18) minutes seems appropriate.

(2–3) The model in Fig. 10.16 describes the problem. We run the simulation for 20,000 min to observe the average number of parts in the machine queue and the average part cycle time. From the Category Overview Report for the Queue, we see that the average queue time is 8.5 min and the average system time is 195.6 with negligible half widths.

(4) Possible benefits from having this model include the ability to explore what would happen to the cycle time if the first-time quality percentage (24%) were improved. This could be used to cost justify system improvement efforts. Also, the

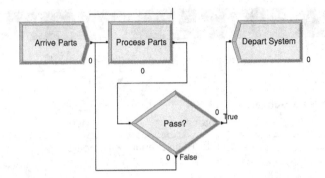

Fig. 10.16 ARENA model for the summary example

benefits from standardizing and/or reducing processing times could be explored. Often, cycle times relate to the lead times sales operations can quote to customers. Reducing cycle times and, therefore, lead times could increase sales. The model could inform inter-departmental meetings with sales and manufacturing engineers trying to decide whether new equipment or training resources are cost justified (Fig. 10.16).

10.7 Problems

1. What module in Arena is used to create a set of resources?
2. What module in Arena is used to destroy entities and record statistics?
3. What does LIFO stand for?
4. What does the Input Analyzer do?
5. What does FIFO stand for?
6. List a standard key performance indicator outputted by ARENA.
7. My wife cuts me off in my exercise after 1 h and I need to spend at least 35 min on the cardio machine. It usually takes about 10 min for things other than cardio but it could take much longer. Develop an ARENA expression that would be a reasonable representation of how long my exercise would take.
8. A coffee shop is trying to understand what would happen if they introduced a third cashier during the lunch rush period. Arrivals are six per minute, service times are approximately exponential with mean 55 s, and the shop loses $1 per minute per customer in lost future sales from long waiting times. Develop an ARENA model to represent this system.
9. Consider the hypothetical measured times 4, 6, 2, 1, 4, 8, 4, 3, 2, 1, 6, 2, 1, 3, and 4. Make a relative frequency histogram, best fit a distribution, and estimate the SSE.
10. Consider the hypothetical measured times 14, 16, 2, 11, 14, 8, 14, 3, 12, 11, 6, 12, 11, 13, and 14. Make a relative frequency histogram, best fit a distribution, and estimate the SSE.

11. Precisely and completely describe the procedure used in the "Fit All" option from the Input Analyzer.
12. Precisely and completely as reasonably possible describe the sum of squared error calculation procedure in ARENA.
13. If the Input Analyzer outputted, "Corresponding $p \ll 0.001$," what would that mean about the fitted distribution?
14. If the Input Analyzer outputted, "Corresponding $p = 0.15$," what would that mean about the fitted distribution?
15. Your store typically has 200 people coming during each 8-h day. The cash registers typically require 4 min per order. How many machines would you recommend and why?
16. Your store typically has 300 people coming during each 13-h day. The cash registers typically require 4 min per order. How many machines would you recommend and why?
17. (Advanced, use a decision module explained in Chap. 11) Develop an ARENA model of an election location in which some voters wait in line, go through registration, and then use booths and optical scanners. At the same time, most voters use ordinary DRE voting machines. Show the module layout needed.
18. (Advanced, use a decision module explained in Chap. 11) Develop an ARENA model of a manufacturing process with a machine, inspection, and optimal rework. Show the module layout needed.
19. Explain two ways of terminating a simulation.
20. Describe the way to send products with different processing times through a manufacturing process.
21. (Challenging) Suppose you would like to model the end of the day as follows. Arrivals after 6 pm are turned away. However, entities in the system at that time are processed to completion. Describe in reasonable detail your approach for terminating the simulated day.
22. (Challenging) Suppose would like to simulate the total amount of money made by a small shop with customers leaving the line if they wait more than 5 min and cashiers costing $25K per year. Describe in reasonable detail your approach for terminating such a system.
23. Suppose arrivals are "very random" in the rush period from 12 to 1 pm with typically 200 people on average coming in that period but with wide variation. Also, assume that there are two cash registers each processing about 1 person per minute with standard deviation about 30 s. If not asking for signatures could shave off 10 s from the average, would it likely affect the customer experience? You do not need to run an ARENA model. Give a flowchart and processing times consistent with the problem statement, i.e., clarify your distributional assumptions for each entity (process, block, etc.) in your chart.
24. Suppose arrivals are "very random" in the rush period from 3 to 5 pm with typically 300 people coming in that period plus or minus. Also, assume that there are three cash registers each processing about 2 people per minute with standard deviation about 30 s. If not asking for signatures could shave off 20 s

from the average, would it likely affect the customer experience? You do not need to run an ARENA model. Give a flowchart and processing times consistent with the problem statement, i.e., clarify your distributional assumptions for each entity (process, block, etc.) in your chart.

Chapter 11
Advanced Modeling with ARENA

Chapter 10 covered the basics of simulation modeling with ARENA. This chapter explores more advanced ARENA processes including those involved with relatively impressive animation and scheduling. An understanding of the material in Chap. 10 is required.

Table 11.1 summarizes the modules used in this chapter, with a brief definition for reference. Some or all of these modules are likely to be needed for industrial projects, which generally require record keeping for specific types of entities using Assign modules. Also, the path that entities typically take through a system depends on conditions such as the type of entity and "Decide" modules. "Record" modules are needed to keep records of specific values, while the "Station" and "Route" modules are critical for setting up animations that look realistic, i.e., do not appear like a flow chart with the modules or blocks shown.

In particular, the "Decide" module provides perhaps the most critically important way to build structure into simulations. Decide can route entities based on a probabilistic condition such as whether or not parts conform to specifications. More commonly, perhaps, it can route based on system conditions. For example, an entity might enter the shorter queue or be assigned to the more productive process. In the latter case, we might designate Process1 and Process2 as the two servers. Then, the entity routing condition in the Decide block could be based on the expression: Process 1.NumberOut <= Process 2.NumberOut. If the condition is true, the entity departs the decide block through the upper exit point, otherwise, the lower point.

In Sect. 11.1 the use of Stations is described. Section 11.2 deals with related animations. In Sect. 11.3, scheduling is described. Section 11.4 illustrates the application of all blocks in a manufacturing example. In Sect. 11.5 three additional examples are selected to reinforce the methods learned.

T. T. Allen, *Introduction to Discrete Event Simulation and Agent-based Modeling*, 161
DOI: 10.1007/978-0-85729-139-4_11, © Springer-Verlag London Limited 2011

Table 11.1 ARENA module descriptions

☐ **Assign**	This module is used for assigning new values to variables, entity attributes, entity types, entity pictures, or other system variables. Multiple assignments can be made with a single Assign module.
◇ **Decide**	This module allows for decision-making processes in the system. It includes options to make decisions based on one or more probabilities (e.g., 75% true; 25% false). Conditions can be based on attribute values, variable values, the entity type, or an expression.
☐ **Record**	This module is used to collect statistics in the simulation model. Various types of observational statistics are available, including time between exits through the module, entity statistics (time, costing, etc.). Tally and Counter sets can also be specified.
▮ **Route**	The Route module transfers an entity to a specified station. A delay time to transfer to the next station may be defined.
▮ **Station**	The Station module defines a station (or a set of stations) corresponding to a physical or logical location where processing occurs. If the Station module defines a station set, it is effectively defining multiple processing locations.

11.1 Stations

ARENA has advanced modules listed in the project bar under basic processes. To access those tools, right click inside the project bar and go to *Template Panel > Attach....* This will open a new window where the user will have the option to select among many other process bars. To use stations, open the *Advanced-Transfers*.tpo. Once open, an Advanced Transfers tab should appear inside the project bar. The advanced transfer modules give the option to use conveyers along with other transportation techniques.

This book focuses on using stations and routes and the example is illustrated in Sect. 11.4. These transportation processes are created in the same fashion as the basic modules were in Chap. 10. Once a station is created, one can change the names and settings using either method explained in Chap. 10 for changing the names and settings of basic processes.

Note that when using a station for transport, one should create a station module followed by a route module (paired together). This is useful for animation purposes as explained in Sect. 11.2. The route module simply tells the entity which station to proceed to. Routes that send the entity to the named station are not connected; rather the entity travels unseen along an invisible path to the station. The benefit is that the paths can then be animated as described in Sect. 11.2.

The basic process for a station route transfer for a given model is as follows: the entity created then travels through the module. It passes through a station to a route. The route module effectively determines which station to go to next. The transfer continues through stations and route modules. Contained inside the route is the travel time specified by the creator.

11.2 Animation

A low-quality animation is built into ARENA basic modules as described in Chap. 10. Its quality is low in the sense that entities travel visually on top of what is essentially a flowchart or workflow diagram. As a result, when the modules in the model window are shown with the moving entities, the impression is not realistic. At the same time, using Station and Route modules, the entities are not shown unless other actions are taken. This section describes the way Stations and Routes permit relatively high quality animation.

Note that ARENA is fully Microsoft® compatible, i.e., a frequent first step in preparing high quality animation is pasting a picture or layout as a backdrop into the model window. For example, Fig. 11.1 shows an ARENA animation overlaid on the floor plan of a clinical research center. The plan was scanned as a .jpeg image and pasted into the model area above the model blocks. As another example, say that one wished to show the path of baseball bats in a bat-making factory. To do that, one could create an image of a bat, paste it into the bat factory layout and then show the path of the bat as it would be using a floor plan image of the factory.

After pasting in the image, e.g., the floor plan or bat, right-click on the shortcut bar, checking that the animate and animate transfer options are selected. If they were not checked before and now are checked, they will appear on the shortcut

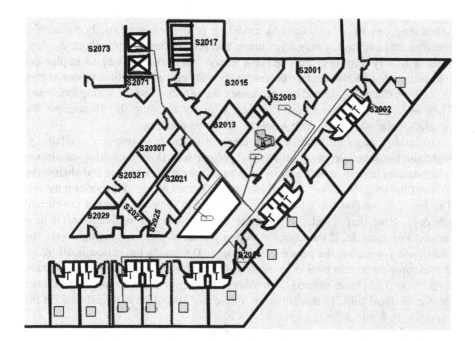

Fig. 11.1 ARENA animation overlaid on the floor plan of a clinical research center

bars. Next, select one of these four ⟨ ⟨ ⟨ ⟨ buttons. The "R" sets a characteristic of a route path object. The "S" specifies a characteristic of a segment of the route. The "D" determines the distance of the route. The "N" specifies a characteristic of a network node or link.

Let us choose, for example, to select the "R" button. A dialog box will appear to set the path by which to travel from station to station. Select OK in the box once the settings have been changed as desired. Then, click where one wishes the path to start. Each additional click creates a kink in the line for turning. Once the desired end of the path is reached, double-click. The path is now set. When you run the ARENA simulation you will see the objects moving along the paths. This method has been used to create the nodes and paths for the clinical research center simulation shown in Fig. 11.1.

Consider also going to *Edit > Entity Pictures* to create or edit the animated pictures so as, for example to make the entities a resemble nurses or patients in the simulation. Another way to do this is to select ENTITY under Basic Processes. Then, change the Initial Picture to the desired picture. If changes later in the model are desired, this can be accomplished using ASSIGN blocks. In the ASSIGN dialog, select "Add Assignments…" and Type "Entity Picture" and then pick the new picture of the entity.

11.3 Scheduling

Scheduling can be a key factor in modeling production systems. For example, consider that employees may have lunch and regular breaks throughout the day. Also, a factory may only be open 8 h a day. An important way to implement scheduling in ARENA centers on resources and their availability. In our voting example in Chap. 10, we had two resources, the registrar and the voting machines. They were both of type "fixed" in their capacity meaning that throughout the simulation the number of each resource remained constant.

To modify this constraint so as to create a schedule of resource availability, highlight the schedule block inside the project bar. Double-clicking inside the schedule area permits schedule customization. One can then type in and change the name of the schedule along with all its settings in the columned drop down menus. The final column contains a button that is pushed to show the graph of the current schedule. Note that scheduling depends on the particular resource unit's time setting. For example, if the resource were a particular person, one might vary the chart from 1, meaning the person is working, to 0 meaning the person is off. Also, the resource could be a bank of workers and/or machines. For example, a company might have 5 machines running 24 h a day. However, it might use only 3 machines during the third shift. In this situation, the setting would be specified as 5 on the chart for 16 h and 3 for the remaining 8 h.

Once the schedule is created, select the resource block in the project bar. In the column type, use the drop down menu to select the "based on" option.

The scheduling column should appear (depending on the version of ARENA). Generally, one can then select the previously created schedule. Note that scheduling of the arrival of entities can also be done in a similar fashion.

When viewing the list of resources in the bottom area, notice that, once scheduling is selected, a scheduling rule column will appear. This column offers three options. Selecting "Wait" means that if a break is about to be taken, resources continue working until they finish processing their current entity before taking a break. "Ignore" means that when resources finish up with an entity, their remaining break is shortened by the time elapsed in completing that entity. The third option is "preempt" which means that when the time for a scheduled break comes, resources take their break immediately and the entity must wait until after the break to be finished.

As an example of scheduling consider the problem of modeling the beginning of the day as follows. Arrivals before 6:30 am begin queuing (say they begin arriving at 6:00 am) and then service begins at 6:30 am. How can this process be modeled? Under the Run menu in Setup you can start the day at 6:00 am. Then, you can simply schedule the resources for your project to have zero capacity until 6:30 am.

Finally, note that the "Read" module can permit the drawing into the models of the times of scheduled arrivals. Scheduled arrivals are perhaps more likely than Poisson arrival process in manufacturing and other sectors.

11.4 Manufacturing Example: Decide, Assign, and Stations

The example in this section illustrates all of the advanced blocks considered in this chapter. A coin-making company employs a multi-step process. The coins are of two types depending on their metallic composition. First, one of the two types of metal arrives at the factory. Workers then cut and stamp the coin from the delivered metal. Next, the coin travels to a plating building on the companies' campus where it is coated with a protective metal. The plating process has a processing time according to a triangular distribution of TRIA(1, 3, 6) in min. Once plated, the coin goes to a third facility where it is inspected. If the unit passes inspection it is then shipped; if it fails inspection it is scrapped. Each inspection requires 5 min. The inspector works an 8-h day with a 1-h break.

The problem is to create an arena model that uses stations and animation. Assign modules are first used to set the two different coin types. The service time for the first type is assumed to be 3; the other's service time is set at 9 (in min). A Decide module is then used to determine whether the coin passes inspection. Assume that the coin has an 80 percent chance of passing.

The next task is to create a facility layout and animate the coins' paths with a relatively high degree of realism.

As a first step, on can generate two Create modules and then connect them to two Assign modules. One of the Assign modules' user forms should have its

Fig. 11.2 Dialog box for the assign module

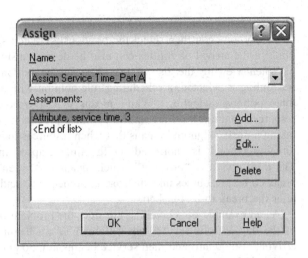

Fig. 11.3 Dialog box for the Decide module

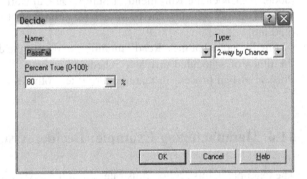

service time attribute set as 3 as shown in Fig. 11.2. The second assign module will have its service time assignment set to 9, instead of 3. This will permit different processing times to be simulated.

Next, we create a cutting process module. Instead of using a delay time like a pseudorandom triangular or uniform (UNIF) number, we use an expression designating the service time inside the expression area. As a result of our assign module, our process time for cutting will change based on which item is being cut. Note, we had assigned a service time of 3 to one coin type and with a service time of 9 to the other. Following the process module, we create a station and a routing module.

The plating area begins with a station that receives the coins from the cutting area. Next, a plating process is created using the TRIA(1, 3, 6) distribution. Following cutting the Route module sends the coin to the inspection area station. Next, we create a decide module based on Fig. 11.3.

Note that the "Type" area of the decide module can be set to: two-way by chance or two-way by condition. Alternatively, the user can choose N-way by chance or N-way by condition with *N* specifying the desired number of exit

Fig. 11.4 Inspectors
schedule

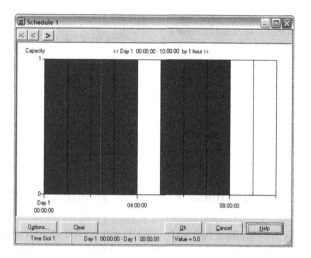

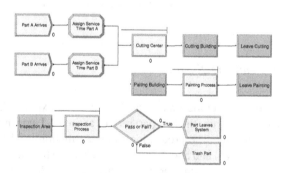

Fig. 11.5 ARENA Model for the coin manufacturing example

points. N-way by chance must, of course, have probabilities summing to 100 percent. Two dispose modules are then created, one for shipping and one for scrap. The schedule for the inspector is then implemented. The inspector schedule inputs should look similar to those in Fig. 11.4.

The entire ARENA simulation layout should then resemble the diagram shown in Fig. 11.5.

Use the animation path-creating features from Sect. 11.2 to create the paths. The result should resemble that shown in Fig. 11.6. When the completed model is run, the results should display a relatively realistic impression of parts moving through the coin facility. Note that the queues can be moved from the blocks in Fig. 11.5 and dragged into the animation part of the model window. It might also be desirable to paste in a facility layout created, for example, by using Microsoft PowerPoint or a drawing package.

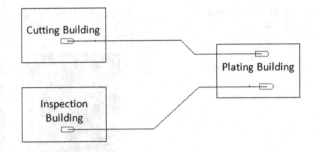

Fig. 11.6 Animations in ARENA showing the nodes and arcs

11.5 Additional Examples: Health Care and Manufacturing

In this section, two additional examples are included that illustrate the application of the "Decide" and "Assign" blocks, which might be considered the most important blocks introduced in this chapter. The results here are based on twenty replications. The extensions to provide half widths based on forty or more replicates are given in the problems at the end of the chapter.

11.5.1 Single Machine System

Parts arrive at a single machine system according to an exponential interarrival distribution with mean 20 min; the first part arrives at time 0. Upon arrival, the parts are processed at a machine. The processing-time distribution is TRIA(11, 16, 18) min. The parts are inspected and about 24% are sent back to the same machine to be reprocessed (same processing time). Run the simulation for 20,000 min to observe the average number of parts in the machine queue and the average part cycle time (time from a part's entry to the system to its exit after, however, many passes through the machine system are required).

In generating the ARENA model, we start with a "Create" module with EXPO(20) interarrivals. Next, we include a "Process Module" with a single resource and using "Seize-Delay-Release" as usual. The "Decide" module is two-way by chance. The False state is then routed back to the process module. Otherwise the routing is to the "Dispose" module. The architecture is given in Fig. 11.7. The module names are changed to improve readability.

We simulate 20 replications and the estimated expected total system time equals 195.6 min. Also, the estimated expected number in queue equals 8.5 items. Normally, we quote "±" half widths but we leave these to problems at the end of the chapter. A model like this could be useful for clarifying, e.g., how improvements in first-time quality (i.e., the factor representing the percentage of parts passing inspection each time) could lead to inventory reductions (i.e., a response which is the expected number in queue).

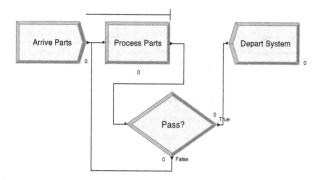

Fig. 11.7 ARENA model of a single server system

11.5.2 Acute-Care Facility

An acute-care facility treats non-emergency patients (cuts, colds, etc.). Patients arrive according to an exponential interarrival-time distribution with a mean of 11 (all times are in min). Upon arrival they check in at a registration desk staffed by a single nurse. Registration times follow a triangular distribution with parameters 6, 10, and 19. After completing registration, they wait for an available examination room, there are three identical such rooms. Data show that 55% of the patients have service times that follow a triangular distribution with parameters 14, 22 and 39. The rest of the patients (45%) have a triangular service time distribution with parameters 24, 36 and 59. Upon completion, patients are sent home. The facility is open 16 h each day.

In generating the ARENA model, we start with a "Create" module using EXPO(11) interarrival times. For complicated models, we usually want to be able to control our measured variables so we start with an "Assign" module to initialize our arrival time. This sets up tracking with a "Record" block at the end. Next, we have the check-in process module. This is followed by a "Decide" module which makes the assignment probabilistically. The "Assign" modules that follow specify the specific service distributions for each type. Then, we terminate with a "Process" module and "Record" and "Dispose" module sequence. The architecture is in Fig. 11.8.

As usual, we base our estimated expected values on 20 replications (or more with each replication representing a single simulated day). The resulting estimated average total patient time in system is 106.4 min. Normally, we would quote "±" half widths on our estimates, but we leave that for the problems at the end of the chapter.

Consider that the model in Fig. 11.8 might be used to explore the effects of factors such as the number of examination rooms and the number of open hours for the clinic on responses including the average total patient time in the system.

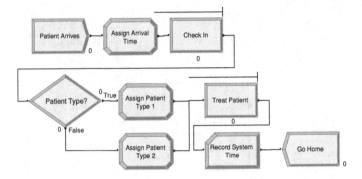

Fig. 11.8 ARENA model of an acute-care facility

11.6 Problems

1. List a module in ARENA useful for keeping track of custom statistics.
2. List a module in ARENA useful for realistic animation of movements of entities.
3. What does UNIF stand for?
4. What does WEIB stand for?
5. What are three types of scheduled failure rules?
6. What are three types of scheduled resource available rules?
7. Explain the IGNORE schedule rule using a graph.
8. Explain the WAIT schedule rule using a graph.
9. Explain the PREEMPT schedule rule using a graph.
10. Which schedule rule would describe workers who immediately leave their processing at their break time?
11. (Challenging) What expression can be used in a "Decide" module to check which of two stations has a shorter queue, to determine which to enter (if it's a tie, they enter station 1)?
12. (Challenging) What expression can be used in a "Decide" module to check whether someone had waited in line more than 60 min?
13. What is the attribute name used in ARENA to denote the current simulation time?
14. Give an ARENA expression for the difference in daily opening times.
15. Consider an ARENA model built that has three different entity types being created and going through processes before being served. How could you use modules to determine: (a) number people of each type moving through the system? And (b) the average cycle time across entity types? (c) Consider an ARENA model of a production system with arrivals coming with inter-arrival times of EXPO(5) min. With probabilities 60, 30, 10% the entities are part type A, part type B, and part type C. Without using any "Decide" modules, how could you assign entity and part type-specific processing times? You do not have to give all of the specifics, but you should clearly outline your reasoning.

16. Consider a pair of faculty generating grant proposals that take 10 days of work on average but whose completion times are highly variable. Half the proposals require both faculty members and the other half require only one of them. Assume that faculty cannot work on more than two proposals at a time and wait doing other activities if necessary. Use simulation to estimate how many proposals the faculty pair can expect to generate in a year.

17. A factory is considering investing in a higher quality robotic device to make its product and comparing that possibility with simply purchasing a second device and running two parallel lines. Currently, 25% of its robot's produce typically needs reworking. Right now, the factory needs to ship only 325 units per 12-h shift but demand is pushing that to 425. The managers want you to help them understand their quality-versus-productivity tradeoff.

   a. Suppose that you time 12 inter-arrivals of parts at the robot. The results are: 90, 95, 90, 90, 98, 92, 93, 94, 91, 88, 89, 91, 92, 87, and 90. Also, suppose that rework is "very random" typically requiring 10 min "plus or minus 10 min" after which the item is re-run through the robot. Also, suppose that the factory managers tell you to assume that the robot operations take a time that is TRIA(80, 85, 90) in seconds based on a past study. Perform input analysis to estimate what you need for simulation and show your results in a flow chart.

   b. Perform simulation with a sufficient number of replications to estimate the number of expected units shipped. Provide an estimate for the half width. Also, display the expected numbers shipped for at least 3 values of the fraction of units needing rework.

   c. Provide decision-support for the company declaring conclusions and characterizing your own uncertainties appropriately.

18. A factory is considering investing in a higher quality robotic device to make its product and comparing that possibility with simply purchasing a second device and running two parallel lines. Currently, 20% of the products coming out of its robot typically need rework. Right now, they need to ship only 320 units per 12 h shift but demand is pushing that to 425. They want you to help them understand their quality versus productivity tradeoff.

   a. Suppose that you time 12 inter-arrivals of parts at the robot. The results are: 80, 85, 90, 90, 98, 92, 93, 94, 91, 88, 89, 91, 92, 87, and 90. Also, suppose that rework is "very random" typically requiring 10 min "plus or minus 10 min" and then the item is re-run through the robot. Also, suppose that they tell you to assume that the robot operations take a time that is TRIA(80, 85, 90) in seconds based on a past study. Perform input analysis to estimate what you need for simulation and show your results in a flow chart.

   b. Perform simulation with a sufficient number of replications to estimate the number of expected units shipped. Provide an estimate for the half width.

Also, display the expected numbers shipped for at least 3 values of the fraction of units needing rework.

c. Provide decision-support for the company declaring conclusions and characterizing your own uncertainties appropriately.

19. What combination of modules (possibly including create, process, assign, ...) would permit modeling the random time it takes for customers to walk through a complicated service system, from the entrance to the exit?

20. What combination of modules would permit modeling of the number of entities of two types currently in a complicated system?

21. Consider an ARENA model of a production system with arrivals coming with inter-arrival times of TRIA(1, 1.5, 2) min. With probabilities 60%, 30%, 10% the entities are part type A, part type B, and part type C. Assume the processing times are EXPO(2), EXPO(2), and EXPO(0.5), respectively and there are two machines with FIFO queuing. Assume a 95%-confidence interval for the average waiting time for all parts over a 10-h day.

22. Consider an ARENA model of a production system with arrivals coming with inter-arrival times of TRIA(1, 1.7, 2) min. With probabilities 65%, 25%, 10% the entities are part type A, part type B, and part type C. Assume the processing times are EXPO(2), EXPO(2.5), and EXPO(0.5), respectively and there are two machines with FIFO queuing. Give a 95% confidence interval for the average waiting time for all parts over a 10-h day.

23. A consultant has recommended that the office from the previous problem not differentiate between customers at the first stage and use a single line with three clerks who can process any customer type. Develop a model of this system, run it for 5,000 min, and compare the results with those from the first system.

24. Develop a simulation model of a doctor's office and estimate the average waiting time.

25. An office that dispenses automotive license plates has divided its customers into categories to level the office workload. Customers arrive and enter one of three lines based on their residence location. Model this arrival activity as three independent arrival streams using an exponential interarrival distribution with mean 10 min for each stream, and an arrival at time 0 for each stream. Each customer type is assigned a single, separate clerk to process the application forms and accept payment, with a separate queue for each. The service time is UNIF(8, 10) min for all customer types. After completion of this step, all customers are sent to a single, second clerk who checks the forms and issues the plates (this clerk serves all three customer types, who merge into a single first-come, first-served queue for this clerk). The service time for this activity is UNIF(2.66, 3.33) min for all customer types. Develop a model of this system and run it for 5,000 min. Provide a confidence interval for the average and maximum time in system for all customer types combined.

26. An office that dispenses automotive license plates has divided its customers into categories to level the office workload. Customers arrive and enter one of

three lines based on their residence location. Model this arrival activity as three independent arrival streams using an exponential interarrival distribution with mean 13 min for each stream, and an arrival at time 0 for each stream. Each customer type is assigned a single, separate clerk to process the application forms and accept payment, with a separate queue for each. The service time is UNIF(8, 11) min for all customer types. After completion of this step, all customers are sent to a single, second clerk who checks the forms and issues the plates (this clerk serves all three customer types, who merge into a single first-come, first-served queue for this clerk). The service time for this activity is UNIF(2.00, 3.33) min for all customer types. Develop a model of this system and run it for 5,000 min. Provide a confidence interval for the average and maximum time in system for all customer types combined.

27. Customers arrive at an order counter with exponential interarrivals with a mean of 10 min; the first customer arrives at time 0. A single clerk accepts and checks their orders and processes payments, taking UNIF(8, 10) min. Upon completion of this activity, orders are randomly assigned to one of two available stock persons (each stock person has a 50% chance of getting any individual assignment) who retrieve the orders for the customers, taking UNIF(16, 20) min. These stock persons only retrieve orders for customers who have been assigned specifically to them. Upon receiving their orders, the customers depart the system. Develop a model of this system and run the simulation for 5,000 min, observing the average and maximum customer time in system.

28. Develop a reasonable model of an automotive manufacturing system and generate a confidence interval for the average number of items shipped even accounting for nonconformities and rework.

29. The following two questions relate to a paper processing center.

    a. Stacks of paper arrive at a trimming process with interarrival times of EXPO(10); all times are in minutes. There are two trimmers, a primary and a secondary. All arrivals are sent to the primary trimmer. If the queue in front of the primary trimmer is shorter than five, the stack of paper enters that queue to wait to be trimmed by the primary trimmer, an operation of duration TRIA(9, 12, 15). If there are already five stacks in the primary queue, the stack is balked to the secondary trimmer (which has an infinite queue capacity) for trimming, of duration TRIA(17, 19, 21). After the primary trimmer has trimmed 25 stacks, it must be shut down for cleaning, which lasts EXPO(30). During this time, the stacks in the queue for the primary trimmer wait for it to become available. Animate and run your simulation for 5,000 min. Collect statistics, by trimmer, for resource utilization, number in queue, and time in queue.

    b. Describe a response and two factors that might be studied using the above simulation.

30. Provide a half width for the average total patient time in system (sojourn time) in the acute care facility in Sect. 11.5. Base your answer on at least 40 or more replicates.

31. Describe at least one way to use ARENA to model a "kanban" system in which previous stations are shutdown when the number waiting at a downstream station gets too high. Use an example if it helps you explain the key elements of your approach.
32. Provide half widths for both the average total time in system and the average number of parts in inventory for the single machine system example in Sect. 11.5. Base your answer on 40 or more replicates.

Chapter 12
Agents and New Directions

In this chapter, computer simulation approaches in addition to discrete event simulation are described. The focus is primarily on agent-based modeling which is defined as the activity of simulating system-wide properties as they derive from the actions and interactions of autonomous entities. This contrasts with system dynamics and other differential equation-based modeling including finite element methods (FEM). In FEMs, e.g., there are relatively few entities and the interplay of physical or cognitive forces dominates.

The degree of autonomy of agents in the models varies greatly. The entities in discrete event simulation generally have low levels of autonomy in that they rarely change state or learn based on model conditions and follow tightly controlled paths. Therefore, discrete event simulation can be viewed as a type of low autonomy agent-based modeling. The "agent-based" modifier applies if the entities commonly change behaviors conditionally (i.e., learn or are influenced) and/or entities move in directions other than along highly restricted paths.

Note that Monte Carlo simulation (Chap. 2) and its alternatives (Chap. 8) are a major part of virtually all types of simulation involving uncertainty including agent-based modeling and stochastic FEM. This makes the basic statistics in Chap. 2, the input analysis in Chap. 3, and the output analysis from Chap. 6 relevant to many types of simulation including agent-based modeling.

The concept of "emergence" is relevant to all the types of computer simulation considered in this chapter. Emergence is the manner of interaction of large numbers of entities and the patterns that arise from these interactions. The associated multiplicity makes the systems complicated enough that simple physics equations cannot accurately predict the "emergent" system properties, are also known as system responses. Discrete event simulation is also primarily concerned with emergent properties.

In practice, discrete event simulation is distinguished by the functioning of the discrete event simulation controller described in Chap. 4 and presented in code in Chap. 11. Another distinguishing feature is the common focus in discrete event simulations on expected waiting times. Still, other responses, including expected

T. T. Allen, *Introduction to Discrete Event Simulation and Agent-based Modeling*, DOI: 10.1007/978-0-85729-139-4_12, © Springer-Verlag London Limited 2011

system profits and throughput, are commonly modeled with discrete event simulation. Variable definitions in ARENA, e.g., permit the modeling of many types of system responses.

Consider the following varied list of emergent properties:

- expected waiting times, downtimes, or throughputs,
- expected number wounded in evacuations or military engagements,
- population size in biological cases, and
- pull-apart force for a snap fit.

Simulation method types differ largely because of the varied emphasis on uncertainty, the attributes of individual entities, and the centrality of physics.

The next section describes six types of simulation including discrete event and agent-based modeling. In Sect. 12.2, the history of agent-based modeling including the relationship with discrete event simulation is described. Section 12.3 describes how influential agent-based modeling software (NetLogo) can produce information about expected waiting times comparable to discrete event simulation code. The return to the voting systems example permits exploration of the similarity and differences between agent-based modeling and discrete event simulation. Section 12.4 discusses new directions for simulation research and practice. Finally noted is the potential contribution of discrete event simulation and agent-based modeling to addressing each of the grand engineering challenges for the twenty-first century described by the National Academy of Engineering.

12.1 Agent-based and Other Types of Simulation

Table 12.1 describes six types of computer simulation related to emergent system response prediction as a tool for system improvement. The list is varied but all methods listed are potentially relevant for informing related decision problems

Table 12.1 Selected types of simulation and how they operate

Simulation type	Engine	Relevance and data sources
Agent-based	Agent rule iteration and Monte Carlo	Particularly relevant for studying incentives and restrictions and based on low-level data
Discrete event simulation	Event controller and Monte Carlo	Particularly relevant for studying production systems and based on low-level data
Forecasting	Empirical modeling including least squares	Particularly useful for predicting new emergent properties based on high-level data
Markov chain models	Linear algebra	Relatively simple and transparent and based on expert opinion and/or high-level data
Systems dynamics models	Differential equation numerical solvers	Particularly relevant for studying the impacts of decisions based on expert opinion
Physics-based	Finite element methods (FEM) numerical solvers	Particularly relevant for designing engineered products and based on low-level data

with predictions of future system responses. Also, all methods listed are potentially helpful in combination in solving system design problems.

A distinction is made with respect to the type of data source that supports the model development. Here, the term "high-level" data is applied to numbers describing emergent properties such as total sales or average waiting times. The reason to define high-level data is the potential for the emergent system response numbers to derive from real physical system, expert opinions, or simulations.

"Low-level" data are numbers, such as service times, that relate to specific subsystems. Chapter 3 describes the systematic collection of low-level data to support discrete event simulation. Low-level data are often critical for the accuracy of high-level predictions. In particular, physical simulations of rigid bodies often have emergent properties that critically depend on every single boundary condition or low-level data point. At the same time, the Theory of Constraints (Chap. 7) implies that many low-level data have no effect on emergent properties since they are part of non-bottleneck subsystems.

Agent-based modeling involves the development of rules for individuals or entities and for the environment or background. The simulation proceeds as individuals execute their actions as allowed by the environment. Iterating through a fixed number of agents might actually make it easier to apply variance reduction techniques (Chap. 8) than to the simulation regulated by a discrete event simulation controller. In discrete event simulation controller operation, the number of random variables used per replication can vary greatly. This can occur, e.g., because individuals can have identities with specific rules tailored to them such as occurs when ARENA assign module makes conditional assignments (Chap. 11). Agent-based modeling also incorporates uncertainty through a stream of pseudo-random numbers as described in Chap. 2. Thus, the difference between agent-based modeling and discrete event simulation is largely in emphasis and, to a lesser extent, in structure. For example, if only a few agents can act at a given time, then the discrete event controller is simply more computationally efficient than iterating through all possibly relevant agents since the event controller goes immediately to the next entity that can act.

Yet, the focus on the individuals and possible learning behaviors make agent-based modeling worthy of consideration. Packages like NetLogo offer relatively less costly ways to model individual-driven systems than discrete event simulation packages like ARENA. ARENA focuses on systems dominated by well-understood, routine phenomena such as material handling in high volume manufacturing. Individualizing entities is difficult because of the many assignments that need to be made and the individuals mainly interact through processes or other blocks. NetLogo focuses on the individual and the associated tactics and constraints related to interactions with other individuals.

Forecasting techniques are widely used to predict the demand for products and for budgeting resources. Forecasting might not be considered a computer simulation method, but it does involve using computers to predict future events and to evaluate hypothetical alternative actions. Forecasting is probably the most widely-used approach among the six. Yet, it is limited by the availability of high-level data.

By contrast, both discrete event simulation and agent-based modeling provide information for cases in which relevant high-level data are scarce but low-level data can be measured or assumed. For example, a manufacturer considering adding a new type of assembly conveyor may have no data about the production capacity taking into account downtime and the detailed production rules of the relevant facility. Also, counties determining the needed number of voting machines may have little or no data about historical average waiting times or other data that could guide estimates of expected waiting times for ballots in future elections at specific locations.

Markov chain models also might not be considered simulation. These are methods based on an assumed set of system states and assumed transition probabilities between states. Here, these methods are designated a form of simulation because they can permit computers to generate an animated set of system trajectories through time.

Markov chain methods are often based on assumptions about transition probabilities and a truncated or simplified list of available system states. Yet, the entire system can be completely described in a concise way, which is often not true for the alternative approaches. This permits the checking of results by others and facilitates consensus building about alternatives.

The final two methods, systems dynamics models and physics-based models, are based on numerical solutions to differential equations. Systems dynamics models are often generated through humanistic processes of eliciting hypothetical forces on the relevant system including feedback and control.

Physics-based models derive from low-level stress-strain and boundary condition data and assumptions about relatively simple physical systems. For example, Fig. 12.1 shows a finite element method (FEM) physics-based simulation of a plastic snap fit being stressed to breaking. An application of design of experiments (DOE, Chap. 5) using FEM simulations of snap tabs generated a highly accurate metamodel to predict pull-apart force. This model informed decision-making and led to dramatically improved snap tabs for an automotive manufacturer (Allen 2010).

Consider the relevance of combinations of the six methods. Such combinations already occur in computer games and animated movies. For example, a popular

Fig. 12.1 Snap tab finite element simulation (FEM) to help inform engineering design

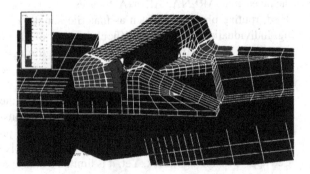

Table 12.2 The spectrum of complexity and validation

	Simple and validated	Complicated and hypothetical
Examples	Manufacturing discrete event simulations and FEM simulations of snap tabs	Disaster evaluation agent simulations and system dynamics models of large economies
Validation	Quantitative with manageable extrapolation	Qualitative, face validity of derived stories from subject matter experts
Roles	Prescriptive	Descriptive, thought provoking

movie can be viewed as a single replicate of a discrete event simulation with specific interactions between entities governed by physical interactions such as violence or limits being placed on movement, e.g., punches and door closings respectively.

Also, the concepts of model complexity and validation are similar in important ways for all the six types of simulation considered here. Table 12.2 summarizes the spectrum of possible models relevant to simulation projects. Models of any type can be relatively simple and closely validated or complicated and subjective. As described in Chap. 2, there is an inevitable "leap of faith" in any attempt to predict the future and extrapolate from historical data. Yet, accuracy in certain predictions is far more likely than in others.

Further, the types of validation attempted can also vary from detailed confirmation using high-level real-world data to the derivation of "stories" from studying simulations. These stories can be studied and critiqued by subject matter experts (SMEs). Are the stories believable or "face valid," or do they have important missing elements that can motivate changes to the underlying simulation models? Relative to the level of complexity and validation is the level of confidence decision-makers can and should have in the model predictions.

At one end of the confidence spectrum, the author is familiar with instances in which discrete event simulations with a high degree of validation together with associated recommendations were used by managers to overhaul major production facility protocols and related purchasing plans. At the other end of the spectrum are thought-provoking simulations based on system dynamics of the possible response of state economies to new regulations. The latter type of simulation might not be trusted to predict accurately the budget for the next year but could lead to surprising insights about the possible side effects of policy decisions.

12.2 The History of Agent-based Simulation

Discrete event simulations usually involve generic entities following rules dictated by system agents or processes such as servers which change their status and cause the entities to wait. That is, the entities themselves neither make decisions nor adapt to changing circumstances. This situation is relevant for many problems in routine systems, but sometimes we would like to represent entities that learn and make choices.

For these applications, many analysts turn to *agent-based simulation,* a relatively new approach made possible by recent advances in computer software. In agent-based models, the agents have decision-making rules along with learning rules or adaptive processes. The focus is on the individuals and pairwise interactions between them. Discrete event simulation packages like ARENA permit the exploration of what is essentially agent-based modeling through the assignment of attributes to entities and rules for specific interactions. At the same time, agent-based simulation packages like NetLogo permit the relatively time-efficient exploration of the implications of individual incentives and conflicts.

Agent-based modeling has roots that some researchers have traced back as far as the 1940s, but serious computerized agent-based modeling began with the Santa Fe Institute's introduction of the SWARM language in the mid-1990s. It expanded somewhat with RePast, in the late 1990s, and became much more widespread with the introduction of NetLogo in the early 2000s.

NetLogo offers open-source (free) software hosted by Northwestern University. The software is described more fully in the next section. Readers can use the tutorials to develop entry-level modeling skills in Net Logo in a few days.

MASON, an even more recent development from George Mason University, provides more tools for specifying geography, is also easy to learn, and has become fairly popular. In mid-2009, Argonne National Laboratory, which developed and supports RePast, released ReLogo, a new interface to import NetLogo models into Repast, to enable modelers to access RePast's larger feature set without completely recoding, and to support comparative modeling exercises.

Samuelson (2000) offers a more thorough overview of the early history of agent-based modeling, especially as applied to studying how organizations work, and Samuelson (2005) and Samuelson and Macal (2006) trace more recent developments. Bonabeau (2002) discusses the potential of agent-based modeling as of the time that its modeling software became widely available.

Three ideas central to agent-based models are:

1. Emergence (often simple models of interactions creating often surprising results),
2. Social agents as objects, and
3. Complexity (often agents model the same subjects as physical models).

That is, the entities in the system are not merely passive objects, but active learning agents that interact with each other. System-level behaviors not obvious to the designer of the simulation emerge from the repeated simulations. The interactions and resulting system phenomena are too complicated to be predicted by straightforward a priori analyses.

Agent-based models, therefore, are particularly useful for assessing when equilibriums are likely to cease to exist, what transient behavior can then be expected, what trigger events are likely to promote stability or instability, and how robust the system is likely to be.

Prominent examples of applications range as far as the spread of epidemics, the threat of biowarfare, the growth and decline of ancient civilizations, social networks, word-of-mouth effects in marketing, supply chain management, large-scale evacuations, and organizational decision-making. Agent-based modeling offers interesting opportunities but also poses challenges. The complexity of the interactions can make both debugging and validation challenging.

12.3 NetLogo

NetLogo is an agent-based model programming environment built on the programming language JAVA and authored by Uri Wilensky. The environment can be downloaded free of charge at http://www.ccl.northwestern.edu/netlogo/. NetLogo is still relatively undeveloped compared with programs such as ARENA, Auto-Mod, or other commercial discrete event simulation software. For example, the NetLogo Model Library and on-line community offerings provide relatively few examples of problems for which an analyst could receive compensation. The freely available examples almost exclusively concern relatively simple, imaginative explorations such as the population of sheep and wolves.

Yet, a few NetLogo models have reportedly been developed on a proprietary basis to model the behavior of crows in evacuations and for a select number of other applications. Commercial discrete event simulation software packages, by comparison, have commercially available modules for thousands of specific versions of material handling equipment and easily permit three dimensional, virtual reality viewing.

The three tutorials available through the Northwestern website provide a helpful way to get started with NetLogo programming. Here, we consider code for a model of a voting machine system. The purposes are to comment on the elements of NetLogo code and to compare the capabilities with those of ARENA and other discrete event simulation software.

Figure 12.2 shows a NetLogo file written by the author to simulate election systems. The figure shows the three tabs: Interface, Information, and Procedures. One of the positive elements of NetLogo is the rapid development of interfaces through the interface tab which can be run almost immediately after being created through drag and drop operations. The Information tab stores the author's comments about the code that the author writes. The Procedures tab is where code is written that is linked to the objects in the interface tab.

Before proceeding to the details of coding, consider that the outputs in the interface provide an approximate estimate of the average or expected number of voters waiting at any given time (around 4) and the expected waiting time (around 2 min). The visual information from the Interface tab generally constitutes the primary quantitative output of simulation. This provides an indication that Net-Logo agent-based modeling is on the relatively qualitative end of the spectrum of computer simulation models described in Sect. 12.2.

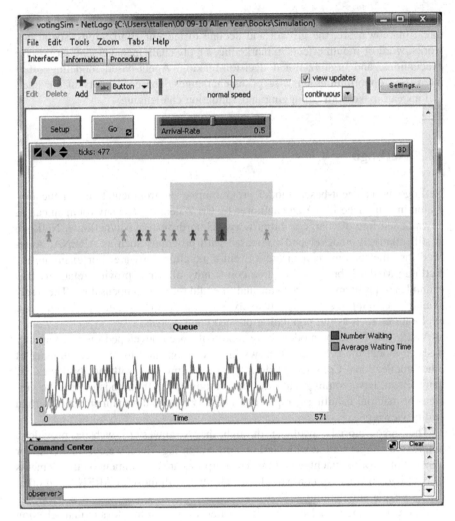

Fig. 12.2 NetLogo simulation of a voting system

Code 12.1 shows the initialization of the NetLogo (.nlogo) program under the Procedures tab in our voting example. It includes the declarations of "global" variables which are equivalent to Public variables in Visual Basic (see Chap. 9), i.e., global variables have values that are accessible to all functions or parts of the code. Text after the ";" is included merely to help clarify the code but not executed. The setup procedure is executed by clicking the setup button in the Interface tab. Functions begin with the word "to" and terminate with the word "end."

Code 12.1 Initialization functions in the Procedures tab for the voting systems NetLogo example

```
globals
[ num-waiting
  accel
  decel
  speed-min
  speed-limit ]

to setup
  clear-all
  ;; Set variable values
    set num-waiting 0
    set accel 0.099
    set decel 0.099
    set speed-min 0
    set speed-limit 1
  ;; Initialize the people
    create-turtles 10
    ask turtles [ set heading 90
    set speed  Arrival-Rate ]
    ask turtles [ setxy random-xcor 0 ]
    set-default-shape turtles "person"
  ;; Prepare the background
    ask patches [ set pcolor white
    if ( pycor < 2 ) and ( pycor > -2 ) [ set pcolor yellow ]
    if (pycor < 5) and (pycor > -5) and (pxcor < 5) and (pxcor > -5)
    [ set pcolor yellow ]
    if (pycor < 2) and (pycor > -1) and (pxcor < 1) and (pxcor > -1)
    [ set pcolor grey ] ]
  end
```

The setup begins with resetting variable values (clear-all) and then defining a few variable values. "Turtles" are the NetLogo designation of agents or individuals. "Ask turtles" calls all agent entities and executes a list of changes to them. In this case, the turtles are voters who are initialized on the visual grid on the x-axis at the zero point but distributed in uniformly pseudo-randomly selected positions (random-xcor). The number of turtles is declared (arbitrarily) to be ten.

Next, the default entity is declared to be a person that affects the appearance, and the orientation is 90° (horizontal). Then, the focus switches to the "patches" or background. "Asking patches" means iterating assignments among background elements. There is an implied looping over all the pixels in the visual grid. In other words, there is iteration over the grid coordinates setting colors and assigning properties. In this example, the only property assigned is the color.

The Code 12.2 continues the code from the Procedures tab. All the code is stored in one file but divided here to fit on book pages. The "turtles-own" designation permits variables to be defined and associated with all specific agents or turtles.

The "go" button on the Interfaces tab (Fig. 12.2) executes the simulation. This is typically done with the "forever" option so that the simulation continues until manually turned off. The go procedure in Code 12.2 starts the voters moving to the voting machine in the middle. The slow-down-person and speed-up-person programs are run to keep the voters from running into each other. The voters or agents leave the right side of the screen and (supposedly new) agents (or turtles or voters) enter the left side. The voting machine stops the agent and causes the line to form. The wait-time variable stores the waiting times.

The plot on the Interface tab links to the plotting function. The plotting function simply tabulates and displays average number of voters waiting and average waiting times. Since there is no replication, the time series display provides a visual approximation to the expected values. Visually waiting for the system to reset, as described in Sect. 8.4 permits some degree of protection from autocorrelation.

The example provides an indication of the value and limitations of agent-based modeling using NetLogo. Advantages of NetLogo compared with ARENA include that NetLogo:

- Is free,
- Has a more easily customized interface (buttons, plots, visualization),
- Permits relatively colorful, intuitive visualization compared with ARENA (which is also two-dimensional in nature), and
- Generates relatively realistic interaction among entities or agents.

With regard to interaction, it is easy in NetLogo to model reactive behavior such as speeding up voting service times if the entity waits in line (preparing for voting) or sees long lines (urgency). At the same time, confidence interval half widths (Chaps. 2, 4, and 10) are not easily generated as in ARENA. Perhaps more importantly, the low-level data used to create the model is more difficult to measure. It relates to acceleration time and movement speeds rather than easily measured arrival rates and service times.

Overall, agent-based simulation permits efficient exploration of individual interactions, incentives, and constraints that is difficult using discrete event simulation. Also, the relatively imprecise nature of NetLogo outputs might be more appropriate for situations in which input data is incomplete. The precision of discrete event simulations can often be misleading considering the level of assumption-making that is often needed to complete the models.

Code 12.2 The go and related functions

```
turtles-own
[ speed    ;; the speed of the voter
  wait-time ;; the amount of time since the last time it moved
]

to go
;; if there -is a person right ahead of you, match its speed then slow down
set num-waiting 0
ask turtles [  set heading  90
  ifelse any? turtles-at 1 0
  [ set speed ([speed] of one-of turtles-at 1 0)
    slow-down-person ]
  ;; otherwise, hurry up
  [ speed-up-person ]
;;; don't slow down below speed minimum or speed up beyond speed limit
  if speed < speed-min [ set speed speed-min ]
  if speed > speed-limit  [ set speed speed-limit ]
  fd speed
  if pcolor = grey
  [ set speed 0 ]
  ifelse speed = 0
  [ set num-waiting num-waiting + 1
    set wait-time wait-time + 1]
  [ set wait-time 0 ] ]
  ;; ask turtles [set label wait-time]
tick
do-plotting
end
```

Code 12.3 The plotting function

```
to do-plotting
set-current-plot "Queue"
set-current-plot-pen "Number Waiting"
plot mean [num-waiting] of turtles
set-current-plot-pen "Average Waiting Time"
plot mean [wait-time] of turtles
end
```

12.4 New Directions

The futures of both discrete event simulation and agent-based modeling are bright. At least four factors will contribute to making them used more widely:

1. Continuing pressures for organizational efficiency,
2. Improved access to low-level data through new sensors and databases,
3. Enhanced visualization capabilities as simulations become more realistic, and
4. Increasing computational efficiencies from faster computers and new simulation-related research similar to methods in Chap. 4 and Chap. 8.

The above trends should increase the usage of methods in areas in which simulation is already being applied. To understand the priorities of simulation-related meetings like the Winter Simulation Conference, stimulating new uses for simulation must also be considered. Such new uses include additional focus on application areas where usage is not yet routine:

- Health care,
- National security,
- Logistics, transportation, and distribution.

There are also new types of technological challenges in addition to application-related challenges. These include combining discrete event simulation with automatic control as described in Kelton et al. (2009). Such combinations promise to influence such diverse areas as robotics and military planning.

Enhanced visualizations will almost surely involve the technological challenge of coupling with physics-based simulations. ARENA is probably on the low end of the spectrum of currently available software with respect to visualization capabilities. The AutoMod model, e.g., in Fig. 12.3, can be shaded and toured easily in what is essentially a limited virtual reality type of interaction. Yet, in the future such tours will almost surely involve interactions with the environment more like a "holodeck" (imaginary entertainment) in "Star Trek" or a tour in "Avatar" with details that enhance believability. Such interactions are important not merely for entertainment but also because they permit opportunities for validation practice, training, and intuition-building.

Additional ways to interact with models besides virtual tours will also likely grow in importance. Such interactions may occur through:

- Gaming and spreadsheet interfaces,
- Multi-fidelity modeling, and
- Directed modeling with high and low-level data.

Agent-based modeling in NetLogo, for example, already includes in the Model Library several examples that offer gaming, including Tetris and more business-relevant exercises. The overall orientation of NetLogo supports gaming. It is relatively easy to develop plausible-seeming models and to interact with them. There is less emphasis on statistical analysis and documentation.

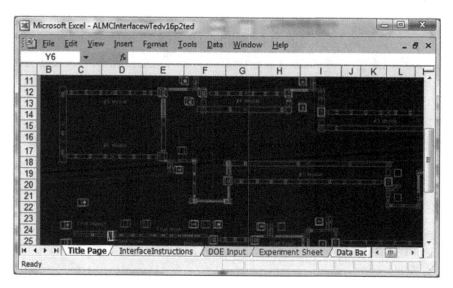

Fig. 12.3 An AutoMod executable embedded in an excel file

This contrasts greatly with older environments that seem relatively serious and uninspiring.

Figures 12.3 and 12.4 show the interface that a major manufacturer currently uses to deploy simulation models. By embedding model executables in Excel, inputs can be changed and models run using only an executable software license. Also, inputs go directly to the spreadsheet where they are integrated into business decision-making and reporting. The software in the interface is available from the author (with fee) and permits the application of methods in Chap. 4 (fractional factorials and simulation optimization) based on "shell" or Windows system calls to program executables using Visual Basic. In this way, users of software with limited output analysis capabilities built-in (e.g., AutoMod) can benefit from up-to-date, computationally-efficient output analysis.

Multiple models of various levels of accuracy or "fidelity" can conceivably be available for the same modeling problem. This situation commonly occurs in finite element method (FEM) physics simulations. High-level or real world response data from the field would generally constitute the highest-level fidelity in these contexts. Data sources, including simulation models, might have different associated costs. Our own research in multi-fidelity optimization is among the first relevant to discrete event simulation (papers first-authored by Huang and Schenk).

Figure 12.5 depicts the outputs from systems of various levels of fidelity and metamodels (Chap. 5) associated with them. The metamodels, which can be regression polynomial or splines such as so-called "Kriging" models, can integrate all inputs and essentially create the most accurate simulation predictions for the highest fidelity system responses. Such methods are still being developed and have only begun to be applied to discrete event simulation systems.

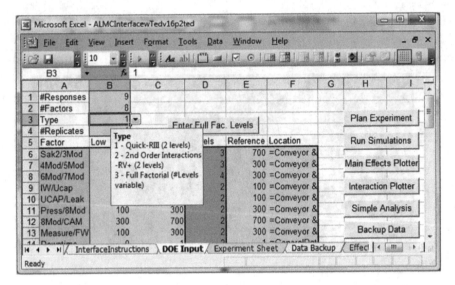

Fig. 12.4 Illustration of the excel interface for an AutoMod simulation executable

Fig. 12.5 Depiction of data of various levels of fidelity and associated metamodels

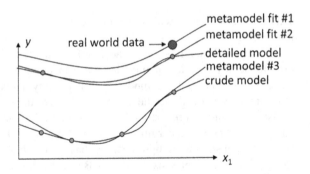

Further enhancements to human-computer interaction with simulation models are also likely to be critical for enhancing the usability of the methods. Currently, the choices that modelers have relate to low-level data collection. If high-level or real world response data become available, an ad hoc adjustment procedure can generally be used to improve the model. This process can become more explicit and open. The initial simulation model can then be made more flexible as well as including a more realistic picture of uncertainties. Then, as information becomes available including through developer or subject matter expert (SME) judgments, the information can be integrated at a high-level. We refer to this research program as "directed modeling" because the human modeler has opportunities to direct development in new ways efficiently.

As a final note, consider the list of "grand challenges" for the twenty-first century developed by the United States National Academy of Engineering. Their list is:

- make solar energy economical
- provide energy from fusion
- develop carbon sequestration methods
- manage the nitrogen cycle
- provide access to clean water
- restore and improve urban infrastructure
- advance health informatics
- engineer better medicines
- reverse-engineer the brain
- personal decision support engines
- prevent nuclear terror
- secure cyberspace
- enhance virtual reality
- advance personalized learning
- engineer the tools of scientific discovery.

It is heartening to consider that discrete event simulation and agent-based modeling can possibly contribute to the accomplishment of every single objective listed. For example, consider that even reverse engineering the brain can involve virtual reality discrete event simulation visualizations in studying how the brain remembers details. Modeling the interactions of agents can inform evolutionary psychology discoveries related to the adaptations of certain brain functions and related behaviors.

Insights from relevant simulations can enable developments while restraining costs and helping engineers, scientists, and business leaders to anticipate and mitigate related side-effects. The methods described in this book and elsewhere can assist in addressing humanity's greatest challenges. Indeed, simulation may well be needed for transparent and fair resource allocation of all types.

12.5 Problems

1. What is an emergent property?
2. What emergent properties are associated with crowds of people?
3. What is "high-level" data and how does it relate to simulation predictions?
4. Which type of simulation is most relevant to predicting the average waiting time of customers in a department store?
5. Which type of simulation is most relevant to predicting the maximum carrying capacity of an elevator?
6. Which type of simulation or modeling is most relevant in industry for predicting the future demand for existing products?
7. Which type of simulation permits concise description and is based on matrix operations or linear algebra?
8. According to the text, what is the relationship of discrete event simulation and agent-based modeling?

9. Describe one method from the chapter relevant for validating simulation models based on a relatively high degree of uncertainty and intended primarily to be thought-provoking only and not prescriptive.

10. Describe a simulation model you might generate and discuss the possibilities for validation. Where does your model fit on the spectrum from simple and validated to complicated and hypothetical?

11. What is NetLogo and who developed it?

12. List two reported professional applications of agent-based modeling.

13. What does the "Settings..." dialog in NetLogo relate to primarily?

14. Experiment with the Schelling housing segregation model in the NetLogo tutorials. How does varying the percentage of people different from me that I'm willing to live with change the resulting housing patterns? Are you surprised by what you see? Can you explain it?

15. Write NetLogo code to change the agents to appear like cars.

16. Experiment with the predator-prey model in the NetLogo tutorials. What ideas can you try here that would be much more difficult to implement in standard discrete-event simulation? How would you modify the model to include the possibility that some prey can communicate danger to each other over a short distance? Over a much longer distance?

17. Write NetLogo code to change the background to red.

18. Consider the end-of-day problem, #11, from Chap. 10. Suppose you want to assume that some people who arrive late know the manager well enough to coax the manager to let them in just after the posted closing time, but only if others who do not know the manager are not present to see this happen. That is, the manager will admit people he or she knows for up to 10 min after the closing time, but only if no one other than the manager knows they are present. How would you model this using agent-based modeling? How would you attempt it using standard discrete-event simulation?

19. What is multi-fidelity modeling?

20. Develop a NetLogo game that is a variant of an existing game. What possible insights can be gained from the resulting application (if any)?

21. Which of the challenges listed in the chapter might best be addressed by discrete event simulation in your opinion?

Chapter 13
Answers to Odd Problems

13.1 Solutions: Chapter 1

1. Input analysis is the process of model formulation, collecting data, and fitting distributions which results in sufficient information to construct a credible simulation model.

3. Output analysis is the use of acceptable simulation or calculated models to generate statistically sound decision support, e.g., using simultaneous intervals or t-testing of outputs.

5. The goal is to develop accurate decision-support information predicting how staffing decisions (factors include the number of staff or each type) would likely affect the expected waiting times of patients (key response) in the med/surg. unit as patient demand continues to increase. The "in scope" area is indicated by the oval in the flowchart below, showing the focus on med/surg. Many details about upstream and downstream units will be simplified or ignored (Fig. 13.1).

7. Responses of interest include monthly expected profits and expected waiting times. Controllable input variables include:

   – Scheduled starting time PM staff (level 1 = 11 am, level 2 = 12:30 pm)
   – Cappuccinos (level 1 = included on the menu, level 2 = not included)
   – Hot sandwich option (level 1 = keep them as an option, level 2 = drop them as an option).

   One might time the interarrivals for only the rush hours (11:45–1:30 pm) and divide food into three types: hot food, other food, and drinks and record the service times for at least 20 of each type.

9. Assume that it takes about 60 h to gather data, make a reasonably complicated simulation model, and validate it (including iteration). This costs about $6,000 which is about as small as any professional consulting project should be to permit sustainability. (Each project typically takes significant time to recruit the

T. T. Allen, *Introduction to Discrete Event Simulation and Agent-based Modeling,*
DOI: 10.1007/978-0-85729-139-4_13, © Springer-Verlag London Limited 2011

Fig. 13.1 A workflow
showing the scope of a
hypothetical project

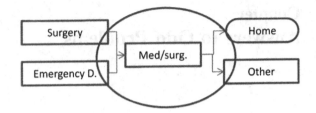

business and set the terms.) It is hard for call centers generally to model the
increases or avoided losses in business for their customers because of their
contribution. Did customer X sign in because they heard of the great customer
service? Yet, we all know intuitively that service is important for bank
competitiveness. Therefore, we will focus only on potential savings relating to
reduced staff expenses. Likely, the call center will be conservative about new
hires and hire only as they feel overloaded (an iterative strategy). Yet, if they do
purchase new software, they may need to make decisions quickly which could
involve delaying or speeding up hiring processes. If through simulation, one of
these processes could be trimmed by only 2 months, the simulation project
would pay for itself with interest $90,000/6 = $15,000.

13.2 Solutions: Chapter 2

1. A number whose value is not known at time of planning by the planner.
3. An LCG is a recursive method to create low quality pseudorandom numbers
   using two sequences with one being $Z_i = \mod(aZ_{i-1} + c, \text{base} = m)$.
5. The sample standard deviation is a summary statistic designed to characterize the spread or dispersion of the generating distribution given by
   $\text{sqrt}[\sum_{i=1,\ldots,n}(X_i - \mu)/(n - 1)]$.
7. Monte Carlo errors are the differences between the sample averages used to
   estimate the mean and the true value of the mean. Generally, one does not
   know the true value so one cannot know the Monte Carlo error exactly.
9. $(1 + 2.4 + 2.0 + 3.5 + 1.4)/5 = 2.06$, $\text{sqrt}\{[(1 - 2.06)^2 + (2.4 - 2.06)^2 + (2.0 - 2.06)^2 + (3.5 - 2.06)^2 + (1.4 - 2.06)^2]/4\} = 0.96$, and (0.96)
   $(t_{0.025,4} = 2.78)/\text{sqrt}(5) = 1.2 \rightarrow [0.86, 2.26]$.
11. $5.6 \pm t_{\alpha/2,n-1}s/n^{-\frac{1}{2}} = (3.4 \text{ to } 7.8)$.
13. $E[X + 3X] = 4E[X] = 4(19) = 76$ which has 0.00000 error. $\text{Var}[X] = E[(X - \mu)^2]$. The three simulated $(X - \mu)^2$ values are: $(10.1 - 19)^2 = 79.21$, $(19.4 - 19)^2 = 0.16$, and $(23 - 19)^2 = 16$. Therefore, the Monte Carlo estimate for $\text{Var}[X] = [(10.1 - 19)^2 + (19.4 - 19)^2 + (23 - 19)^2]/3 = 31.79$. The
   sample standard deviation is 41.8, $t_{0.025,2} = 4.3$ so the half width is
   $\pm(41.8)(4.3)/\text{sqrt}(3) = \pm104$.
15. $E[2X] = 2[X] = 2(22.0) = 44.0$ with 0.0 error (from the definition of
   expected values). Using Monte Carlo simulation or the central limit
   theorem we have: $E[X^2] = (9.1^2 + 20.3^2 + 19.4^2 + 23.0^2)/4 = (82.8 +$

412.1 + 376.4 + 529.0)/4 = 350.1   with   SD = 189.7.   Using a 95% confidence interval, (assuming normality) gives an estimate of 350.1 ± 301.9.

17. $E[X] = (a + b + m)/3 = 5.0$ and, in English, the assumption means that the time is unknown at the current time but it must be >2 h and <10 h with the most likely value equal to 5 h.

$$F^{-1}(y/a, m, b, n) = \begin{cases} a + \sqrt{y(m-a)(b-a)}, & \text{for } 0 \leq y \leq \frac{m-a}{b-a} \\ b - \sqrt{(1-y)(b-m)(b-a)}, & \text{for } \frac{m-a}{b-a} \leq 1 \end{cases}$$

19. Plugging in 0.8, 0.3, and 0.5 for y above gives 6.65, 3.7, and 4.7, repectively.

13.3 Solutions: Chapter 3

1. It is a function that ranges from 0 to 1, increasing at each point in a data set, $X_1,...,X_n$. It is given by $F_n(x) = \sum_{i=1,...,n} Count(X_i \leq x)/n$.
3. There is arbitrariness in determining the number and width of bins in a relative frequency histogram. This affects the sum of squared error (SSE) calculations. Also, in general to achieve subjective believability it is often desirable to insert ad hoc constraints into the SSE optimization or curve fitting process.
5. This was already almost a relative frequency histogram. The only difference was changing the axis ($\div 20$). By eye, the sample mean is around 20 and the sample standard deviation is around 5. More precisely (not needed) mean $\approx \sum_{i=1,...,5} X_i rf(i) = 19.4$. Sample std. deviation $\approx$ sqrt$[\sum_{i=1,...,5} (X_i - 19.4) rf(i)] = 5.4$ (Fig. 13.2).
7. Triangular with $a = 8$, $m = 17.5$, and $b = 35$ seems reasonable. The estimated SSE is $(0.06)^2 + (0.03)^2 + (-0.15)^2 + (-0.15)^2 + (0.09)^2 = 0.06$.
9. The missing bar should have zero height so that all bars sum to 1.0. Relative frequency histograms are typically used (e.g., in the Input Analyzer) to estimate the sum of squares error in fitting given distributions and for comparing distributions.

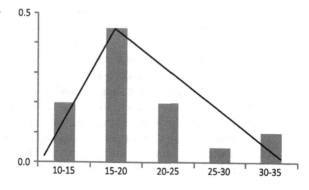

Fig. 13.2 Relative frequency histogram

Fig. 13.3 Relative frequency histogram

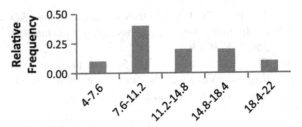

11. The plot shows the cumulative distribution function for the exponential distribution and the empirical cumulative distribution for ten data points. The largest difference between the two is the KS distance used to provide evidence, in some cases, that certain distributions are not a good fit for a given dataset.
13. A relative frequency histogram of the data is below (Fig. 13.3).
15. The exponential would not fit because its density is concentrated near zero.

13.4 Solutions: Chapter 4

1. Independent identically distributed.
3. If $X_1,\ldots,X_n$ are IID, then the sample mean becomes approximately normally distributed with mean given by the expected value and error dev. shrinking as $n^{-\frac{1}{2}}$.
5. The exponential distribution is "very random" or it has a high coefficient of variation. This means that the expiration of electrical components is very unpredictable. A technical term that is often used to describe the exponential is "memoryless" because the instantaneous chance of expiration does not depend on how old the item is.
7. The inverse cumulative distribution for the exponential is $F^{-1}(u) = -\ln(1 - u)/$ (λ). Plugging in the pseudorandom numbers gives pseudorandom exponential numbers 1.78, 4.58, 0.53, and 8.05 respectively. The sample average is 3.73 which is our Monte Carlo estimate for the mean. The true mean for the exponential is λ^{-1} or 5.0. This means that the Monte Carlo error is 1.27.
9. (1) The values are not IID since they are correlated and thus not independent. Therefore, there is no reason to suspect that the sample averages would be characterized by a normal distribution. Thus, in output analysis, we cannot draw conclusions based on normality.
   (2) Again, there is no reason to suspect that the individual values are normally distributed since they do not represent batch averages of IID random variables. Therefore, the sample standard deviation of these non-normally distributed numbers would likely not accurately characterize the standard deviation of the waiting time distribution. Confidence intervals using standard formulas might give misleading results.
11. As the batch size goes to infinity, the sample averages become normally distributed. This assumes that the values in each batch are IID. (IID values can

be achieved by having the replicates be entire simulations of the system in question.)

13. The relative frequency histogram indicates that the batch size is insufficient such that normality has not been achieved. Additional replications and a larger batch size are needed for normally distributed batch averages to be achieved. Then, the sample standard deviations of the means will foster an accurate confidence interval.

15. Generally, the push system would yield a higher throughput because machines would never be starved. However, since the units proposed might not be demanded, the finished goods inventory would likely also be higher for the push system. This analysis is simplified because the inventory in push systems can have many harmful effects. Also, after some time period the push system would not be permitted to continue generating units not being demanded by customers.

13.5 Solutions: Chapter 5

1. Evaluations are noisy so we might find a good solution and lose it. Also, the optimization method not only needs to select which system alternatives to evaluate, but also how many samples for each valuation.

3. Simultaneous intervals are always wider than individual intervals. This follows because we might apply them to make many judgments of significance and we desire to limit our chance of being incorrect in any of these many possible declarations, i.e., they are conservative.

5. $\alpha = \alpha_0/3 = 0.01667$ from the Bonferroni bound. Keeping $1 - \alpha$ intervals, then the overall probability of not making a Type I error is $>1 - \alpha_0 \cdot t_{1-0.01667/2,9} = 2.93$. Therefore half widths are ± 2.4, ± 3.8, and ± 3.3 (Fig. 13.4).

7. $\alpha_{\text{individual}} = 0.05/10 = 0.005$ would guarantee at least a 0.05 probability that all ten system means will be in their corresponding simultaneous confidence intervals.

Fig. 13.4 Simultaneous intervals for the performance of three systems

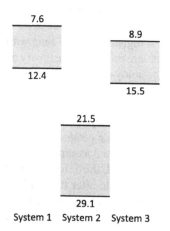

9. An indifference parameter (δ) is a mean or expected value difference that we are comfortable in losing. In other words, if we terminate and do not have the best solution, we are satisfied as long as we have a solution with true mean within this number of the actual best.

11. If we set $\delta = 0.0$, the second stage sample sizes would be infinite. Then, the method would be unusable.

13. The variable sample sizes enter in Step 4 of the second stage of the indifference zone selection method. Systems with high variances are generally allocated the most samples. Hopefully, the worst systems have already been eliminated through the subset selection in Step 2.

15. Alexopoulos (2006) focuses on reviewing methods for extracting output from steady state simulations.

13.6 Solutions: Chapter 6

1. With a complicated model and predictions for situations not covered in historical data, it is particularly important to derive independent confirmation that simulation models are reasonably accurate and trustworthy, i.e., to perform model validation. Queuing theory provides ball park estimates that simulation outputs can be compared with.

3. 12.

5. $p = 0.875$ and $W^q = 38.1$ min.

7. $p = 0.889$ and $W^q = 39.2$ min.

13.7 Solutions: Chapter 7

1. The store manager might be considering purchasing a new oven. He or she might direct the simulation team to study system design alternatives relating to possible oven purchases. Yet, theory of constraints might suggest that alleviating the cash register bottleneck is far more critical. This could occur because the lines at the register are very long while those waiting for food after paying it at the register are minimal. The theory of constraints would then suggest using the simulation to study system design alternatives related to cash register factors are most relevant. Such factors could include increasing the number of cash registers, making sure that register operators do not perform additional operations such as getting drinks, and/or shaving service times by not requiring credit card signatures.

3. The lean producer would toast two pieces of bread, put on peanut butter, put on jelly, and assemble the first sandwich. The, he or she would do the same to the second, third, fourth, and fifth delivering each to the family member as soon as possible after the sandwich was made.

5. They can extend the election period beyond Election Day. Also, they can offload ballot initiatives that might be voted on in November onto other more

minor elections during the rest of the year. In the 2008 November election, some locations in Ohio voted on 24 races and 19 ballot initiatives. This meant that some voters required more than 20 min to vote even after reaching the direct recording equipment (DRE). (Of course, many other states and countries have much shorter ballots requiring voting times less than thirty-seconds.) It is likely true that some of the Ohio races and initiatives could have been voted on during relatively minor elections to level demand.
7. The decision about how many kanban cards are needed might be supported by simulation. Simulation can help managers determine possible tradeoffs between down-time and work in process inventory (WIP) costs or expected lead times.

13.8 Solutions: Chapter 8

1. Simulation is artistic in the sense that the level of detail and realism is largely subjective. As computational power increases so does the temptation to make the simulation increasingly realistic. Also, as simulations become more realistic, decision-makers will generally place more trust in their results. As a result, the need for speeding up simulations is roughly constant or possibly increasing.
3. Each variance-reduction technique is complicated, has a potential issue related to bias in results, a need to anticipate the sample size, or other issue. As a result, simple Monte Carlo based on pseudorandom numbers continues to be viable and dominate the commercial marketplace. With additional results, it is possible that types of variance-reduction techniques or quasi-Monte Carlo will begin to dominate the practice world.
5. No, some types of variance-reduction techniques change the structure of the estimation problem. Instead of simply substituting alternatives for pseudorandom U[0,1] numbers and using the inverse cumulative and controller, they can change the entire approach. For example, importance sampling requires identifying somewhat arbitrarily an alternative distribution with certain properties and basing much of the mechanics on this selected distribution.
7. The estimate is 7.8 min based on Table 13.1.
9. Intuitively, the local situation created in time by specific short or long waits has a decreasing effect as time increases. Effectively, "time heals all wounds" and

Table 13.1 Results for calculating average time estimate

Perm. #1 $(P_{i,1})$	Perm. #2 $(P_{i,2})$	DS $U_{i,1}$ var. 1	DS $U_{i,2}$ var. 2	Registration time	Voting time	Sum (min)
1	4	0.1000	0.7000	0.229	7.551	7.780
2	2	0.3000	0.3000	0.472	6.258	6.731
3	1	0.5000	0.1000	0.754	5.707	6.461
4	5	0.7000	0.9000	1.100	8.586	9.686
5	3	0.9000	0.5000	1.603	6.838	8.441

offers a reset. As the interval between successive samples increases, each observation becomes increasingly like a sample from a random walk and the IID assumption becomes increasingly reasonable.

13.9 Solutions: Chapter 9

1. Three advantages are: (1) there is little or no licensing fee, (2) there is an opportunity to close-couple results with other software being used for accounting and/or optimization, and (3) the resulting code might be more computationally efficient than corresponding commercial codes which are often built with an emphasis on visualization over scientific programming runtime economy.
3. Yes, depending on the operating system and current availability, Microsoft® offers "Express Edition" compilers downloadable in some or all cases with no charge.
5. Flat, Float, Long, String[2], Integer.
7. The following was derived by placing first a number in cell B1. Next, after recording the macro the code was viewed using the Visual Basic Editor.

Code 13.1 Illustration of a simple subroutine

```
Sub Macro1()
    ActiveCell.FormulaR1C1 = "=R[-1]C/27"
    Range("B3").Select
End Sub
```

9. The value 5.18737751763962 was derived using the following code which includes an automatic conversion to double.

Code 13.2 Illustration of a simple subroutine for summation calculation

```
Sub Macro1()
Dim sum As Double, iIndex As Long
sum = 0
For iIndex = 1 To 100
sum = sum + 1 / iIndex
Next iIndex
Sheet1.Cells(1, 1) = sum
End Sub
```

11. The value 288 was derived using the following code.

Code 13.3 Illustration of a While-Wend construction for summation calculation

```
Sub Macro1()
Dim sum As Long, iIndex As Long
sum = 0
iIndex = 1
While iIndex ^ iIndex < 1000
sum = sum + iIndex ^ iIndex
iIndex = iIndex + 1
Wend
Sheet1.Cells(1, 1) = sum
End Sub
```

13. The following code selects a range of five cells and colors them yellow.

Code 13.4 Illustration excel cell coloration

```
Range(Cells(1, 1), Cells(1, 5)).Select
   With Selection.Interior
   .ColorIndex = 6
   .Pattern = xlSolid
   End With
```

15. The following code gives: 2.718281828, 7.389056099, 20.08553692, 54.59815003, i = 5, 403.4287935, 1096.633158, 2980.957987, 8103.083928, 22026.46579.

Code 13.5 Illustration of an If–Then construction

```
Sub Macro1()
Dim iIndex As Long
For iIndex = 1 To 10
If (iIndex = 5) Then
Sheet1.Cells(iIndex, 1) = "i=5"
Else
Sheet1.Cells(iIndex, 1) = Exp(iIndex)
End If
Next iIndex
End Sub
```

17. The following code gives the same sequence as the solution to problem 15.

Code 13.6 Illustration of an If–Then-else construction

```
Sub Macro1()
Dim iIndex As Long
iIndex = 1
While iIndex <= 10
If (iIndex = 5) Then
Sheet1.Cells(iIndex, 1) = "i=5"
Else
Sheet1.Cells(iIndex, 1) = Exp(iIndex)
End If
iIndex = iIndex + 1
Wend
End Sub
```

19. The number 7.80 derives from the following code.

Code 13.7 Illustration of a for loop random variable-based simulation

```
Sub Macro1()
Dim iIndex As Long, sum As Double, nSimulations As Long
nSimulations = 10000
sum = 0
For iIndex = 1 To nSimulations
sum = sum + (-2 * Log(1 - Rnd())) ^ 2
Next iIndex
Sheet1.Cells(1, 1) = sum / nSimulations
End Sub
```

21. The number 508.4 derives from the following code.

Code 13.8 Illustration of a call to an application work sheet function

```
Sub Macro1()
Dim iIndex As Long, sum As Double, nSimulations As Long
nSimulations = 10000
sum = 0
For iIndex = 1 To nSimulations
sum = sum + Application.WorksheetFunction.LogInv(Rnd(), 2, 3)
Next iIndex
Sheet1.Cells(1, 1) = sum / nSimulations
End Sub
```

13.10 Solutions: Chapter 10

1. Set from the Basic Process Panel, type: resource.
3. Last In First Out.
5. First In First Out.
7. Cardio + Other things < 1 h.
9. If one chose to use the ARENA input analysis with fit all function the results are as follows (Fig. 13.5).
11. The procedure creates a relative frequency histogram of the data using an undisclosed routine to select the number of bins. Then, all distributions are fitted and a distribution is selected that minimizes the sum of squared estimate errors, i.e., the differences between the bin fractions and the distribution predictions for the bin means for the best fit distribution of each type.
13. It means that the KS test has proven, with alpha <0.001 that the data did not derive from the fitted distribution function. Likely, the fitted distribution should not be used and an alternative distribution should be found. Factors: Number of primary and secondary trimmers and/or the engineers might be investigating way to reduce service times. Then, factors might be the top end service times for the primary and secondary trimmers. Response: This could be the cycle time, the cost of the WIP, profits etc.
15. This can be answered using various methods, including creating a simple ARENA model and viewing the output. Or one could use simple logic. For instance, 200 people arriving every 8 h is equivalent to one person arriving every 2.4 min. Knowing that it takes 4 min to check a person out, we then know we need 2 machines because we cannot have half or part of a machine. As a result we have 4.8 min per check out which is more time than the needed 4 min (Fig. 13.6).
19. (1) By specifying the replication length
    (2) By specifying a condition

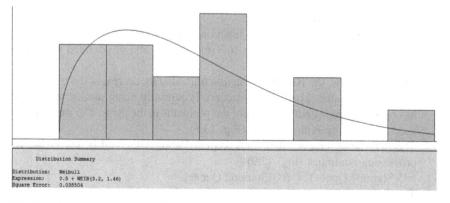

Fig. 13.5 Input analyzer relative frequency histogram and distribution fit

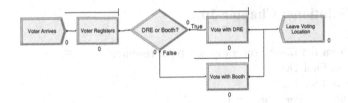

Fig. 13.6 An ARENA model for an election system

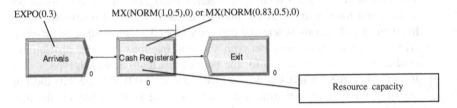

Fig. 13.7 An ARENA formulation for a simple cash register system

21. In your Max Arrivals: box of the create module put a variable. Then, use a logical create, assign, and dispose sequence to change the variable to 0 at the appropriate time. This cuts off arrivals. Then, in the Run → Setup → Replication Parameters panel terminate by expression with the expression being TNOW ≥ 600 && TotalWIP == 0, where TotalWIP records your work in process and 600 refers to the length of the day from start until the time when new calls are not accepted.

23. Fig. 13.7.

13.11 Solutions: Chapter 11

1. The "Record" module.

3. The uniform probability density function or distribution.

5. Ignore, Wait, and Preempt and also Time, Count, and pre-defined state both are acceptable.

7. As shown below, the IGNORE function immediately decreases the resource of an entity no matter whether the resource is currently being used or not. (This has been used repeatedly in most of the problems in the class. The same rule is used for resource failures also.) (Fig. 13.8)

9. The item being processed is not completed and, at the end of the break, its processing continues (Fig. 13.9).

11. NQ(Station1.Queue) ≤ NQ(Station2.Queue).

13. TNOW.

15. Answers follow:

Fig. 13.8 Illustration of the IGNORE rule

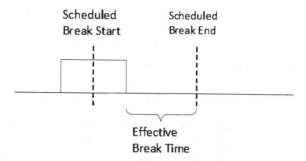

Fig. 13.9 Illustration of the PREEMPT rule

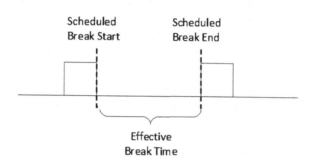

Fig. 13.10 Workflow for a robotic cell

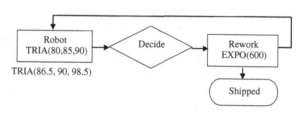

a. For the number people of each type through system—you could assign entity type either in the create module or using an assign module after the create module. Then you could create a set of entity types and put a record module before the dispose node, recording a type count and checking record into set. Alternatively, you could not create a set and before the dispose node, use a three-way conditional decide node to split the types out and then have three record modules each recording a count and named according to their part type (Fig. 13.10).

b. For the average cycle time across entity types, you could use an assign right after the create nodes to assign TNOW as the arrival time. Then, right before the dispose node, you could have a record module with type Time Interval and choose your arrival time attribute as the value.

c. There could be a single EXPO(5) create node. Next, there could be a single Assign module, which assigns part index using a discrete probability distribution like (0.6, 1, 0.9, 2, 1.0, 3) to assign part indexes. In the entity module, you could define your three entity types as type A, type B, and type C in that order. You could also define processing time as an expression with 3 rows. Each row would be the processing time for part A, B, C, respectively. The assign module will then assign the entity type and processing time by using part index in the entity type and processing time arrays.

17. The following models relate to investigating the productivity tradeoff.

   a. The inter-arrival times seem to be bounded with a highest number near the low end. This seems consistent with a triangular distribution. Using the Input Analyzer results in TRIA(86.5, 90, 98.5) which has competitive SSE with alternative distributions and is not ruled out by the KS test.

   b. With 20 replications, there appears to be no issue with correlation or insufficiency. This is possible with fewer but as 20 replications takes only a few seconds to run, it is likely sufficient.

% rework	Expected number of units shipped
25	379.5 ± 6 (half width)
15	431.1 ± 4 (half width)
5	469.4 ± 0.8 (half width)

   c. It seems likely that a robot having a rework rate below 15% might be as valuable as having a second rework line in the sense that meeting the production quota (over 50% of the time) would be possible. Yet, there are a number of issues that might be investigated further. The details of the rework process are critical in this capacity related decision-making. 10 ± 10 min is probably not enough information to make an important decision. Also, it might be worthwhile to investigate what fraction of the days the desired shipping targets can be achieved. Two independent systems are more reliable generally, even while they have much higher operating cost than a single system.

19. Having an Entrance Station, Route block, and a Service Station with the Route associated with a randomly distributed service time meets the specifications of the problem. Alternatively, we can use a process and delay module combination.

21. The ARENA model at right constitutes one representation consistent with the problem description (Fig. 13.11).
    The half width given by ARENA corresponds to a 95% confidence interval. With 20 replications the confidence interval for the expected waiting time is 0.3336 ± 0.07 min.

23. The following ARENA model results in average waiting time estimates (Fig. 13.12).

25. An ARENA model for the license plate dispensing system is shown (Fig. 13.13).

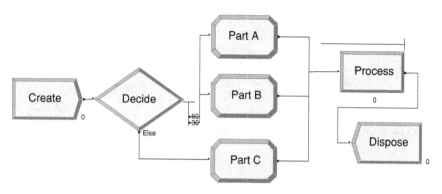

Fig. 13.11 Configuration of ARENA model including Decide and Define blocks

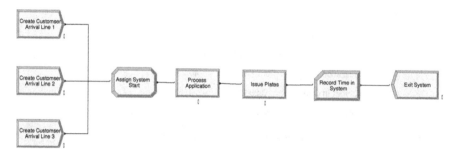

Fig. 13.12 ARENA models and outputs for application processing problem

RESULTS

	Separate Queues	Merged Queue
Average System Time	48.5454	26.5596
Maximum System Time	168.64	67.4135

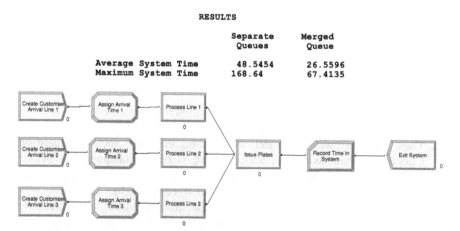

Fig. 13.13 Configuration of ARENA model concerning license plate processing

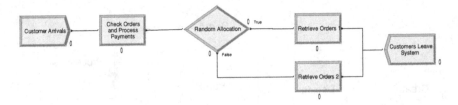

Fig. 13.14 Configuration of ARENA model concerning order retrievel

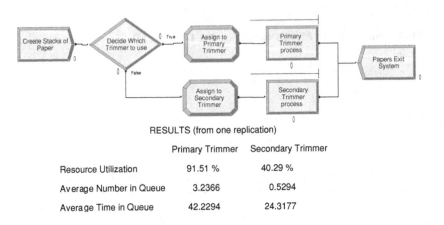

RESULTS (from one replication)

	Primary Trimmer	Secondary Trimmer
Resource Utilization	91.51 %	40.29 %
Average Number in Queue	3.2366	0.5294
Average Time in Queue	42.2294	24.3177

Fig. 13.15 Configuration of ARENA model concerning paper trimming

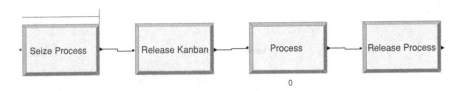

Fig. 13.16 Blocks in ARENA using the advanced Seize and Release modules

27. The following shows an ARENA model of order retrieval (Fig. 13.14).
29. The key is to use a count-based failure to shut down the trimmer (Fig. 13.15).
31. There are many ways to model kanban using ARENA. Some are based on modeling kanbans as entities and others are based on modeling kanbans as a resource. Below shows the key steps in modeling kanbans as resources using the advanced process seize and release blocks (Fig. 13.16).
    In general, separating seize and release operations and using a "Decide" by condition approach can provide approaches with reasonable computational efficiency. More approaches are described in Treadwell and Herrman (2005).

13.12 Solutions: Chapter 12

1. A high-level system output derived from the interaction of multiple entities.
3. High-level data are numbers describing emergent properties and that derive from simulation predictions, measured system responses, or other predictions.
5. Physics-based.
7. Markov chain models.
9. Developing a story from a single simulation run and receiving feedback from a subject matter expert (SME) about its realism provides some level of validation.
11. NetLogo is a development environment for agent-based modeling developed by Uri Wilenski of Northwestern University.
13. The "Settings…" dialog is primarily concerned with the "World" or the set of patches to be viewed.
15. set-default-shape turtles "car".
17. ask patches [set color red].
19. Multi-fidelity model is the development of metamodels in a way that integrates responses from more than a single simulation or system representation.
21. Engineering the tools of scientific discovery could involve simulation because the developer can more immediately perceive development costs and access issues.

References

Adan I, Resing J (2002) Queueing theory. Technical report. Department of Maths and Computer Science, Eindhoven University of Technology, The Netherlands

Agent-based Model Wikipedia http://en.wikipedia.org/wiki/Agent-based_model

Alexopoulos C (2006) A comprehensive review of methods for simulation output analysis. In: Perrone LF, Wieland FP, Liu J, Lawson BG, Nicol DM, Fujimoto RM (eds) Proceedings of the 2006 winter simulation conference

Allen TT (2010) Introduction to engineering statistics and six sigma: statistical process control and design of experiments and systems, 2nd edn. Springer, London

Allen TT, Bernshteyn M (2008) Helping Franklin county vote in 2008: Waiting line report to Michael Stinziano and Matthew Damschroder and The Franklin County Board of Elections http://vote.franklincountyohio.gov/assets/pdf/press-releases/PR-07302008.pdf

Allen TT, Bernshteyn M (2006) Mitigating voter wait times, Chance magazine. The American Statistical Association, Autumn issue

Bechhofer RE, Santner TJ, Goldsman D (1995) Design and analysis of experiments for statistical selection, screening and multiple comparisons. Wiley, New York

Bonabeau E (2002) Agent-based modeling: methods and techniques for simulating human systems. Proc Natl Acad Sci 99(3):7280–7287

Carley KM (2009) Computational Modeling for reasoning about the social behavior of humans. Comput and Math Organ Theor 15(1):47–59

Carley KM, Fridsma D, Casman E, Yahja A, Altman N, Chen LC, Kaminsky B, Nave D (2006) Biowar: scalable agent-based model of bioattacks. IEEE T Syst Man Cy A 36(v2):252–265

Caflisch RE (1998) Monte Carlo and Quasi-Monte Carlo Methods. Acta Numerica v7, Cambridge University Press, New York

Chen EJ, Kelton WD (2000) An enhanced two-stage selection procedure. In: Joines JA, Barton R, Fishwick P, and Kang K (eds) Proceedings of the 2000 Winter Simulation Conference. Institute of Electrical and Electronics Engineers, New Jersey 727–735

Czeck M, Witkowski M, Williams EJ (2007) Simulation improves patient flow and productivity at a dental clinic. In: Zelinka I, Oplatková Z, Orsoni A (eds) Proceedings 21st European Conference on modelling and simulation

Ehrlichman SMT, Henderson SG (2008) Comparing two systems: beyond common random numbers. In: Mason SJ, Hill RR, Mönch L, Rose O, Jefferson T, Fowler JW (eds) Proceedings of the 2008 winter simulation conference

Epstein JM, Axtell R (1996) Growing artificial societies: social science from the bottom up. MIT Press/Brookings Institution, Cambridge

Fu MC, Glover FW, April J (2005) Simulation optimization: a review, new developments, and applications. In: Kuhl ME, Steiger NM, Armstrong FB, Joines JA (eds) Proceedings of the 2005 winter simulation conference

Gilbert N, Troitzsch K (2005) Simulation for the social scientist, 2nd edn. Open University Press, UK

Goldratt EM (2004) The goal: a process of on-going improvement, 3rd edn. North River Press, USA

Goldsman D, Nelson BL, Opicka T, Pritsker AAB (1999) A ranking and selection project: experiences from a university-industry collaboration. In: Farrington PA, Nembhard HB, Sturrock DT, and Evans GW (eds) Proceedings of the 1999 winter simulation conference. Institute of Electrical and Electronics Engineers. New Jersey, 83–92

Holland JH (1995) Hidden order: how adaptation builds complexity reading. Addison-Wesley, USA

Huang D, Allen TT, Notz W, Miller RA (2006) Sequential kriging optimization using variable fidelity data. Struct Multidiscip O 32(v5):369–382

Huang D, Allen TT, Notz W, Zheng N (2006) Global optimization of stochastic black-box systems via sequential kriging meta-models. J Global Optim 34(v3):427–440

Kelton WD, Sadowski RP, Sturrock DT (2009) Simulation with ARENA. 5/e edn. McGraw-Hill, New York

Keynes JM (1923) A tract on monetary reform. rep. 2000 Prometheus Books, London

Koenig LW, Law AM (1985) A procedure for selecting a subset of size m containing the l best of k independent normal populations, with applications to simulation. Commun Stat-Simul C 14:719–734

Kolesar PJ, and Green LV (1998) Insights on service system design from a normal approximation to Erlang's delay formula, Prod Oper Manag Vol. 7(3):282-293

Law AM (2008) How to build valid and credible simulation models. In: Mason SJ, Hill RR, Monch L, Rose O, Jefferson T, Fowler JW (eds) Proceedings of the 2008 winter simulation conference

Lemieux C, L'Ecuyer R (2000) Using lattice rules for variance reduction in simulation. In: Joines JA, Barton RR, Kang K, Fishwick PA (eds) Proceedings of the 2000 winter simulation conference

Liker J (2005) The Toyota way fieldbook: a practical guide for implementing Toyota's 4Ps. McGraw-Hill, New York

Medina RA, Juarez HA, Vazquez A, Gonzalez RA (2008) Mexican Public Hospitals: a model for improving emergency room waiting times. In: Mason S, Hill R, Mönch L, Rose O (eds) Proceedings of the 2008 winter simulation conference

Mesquita MA, Hernandez AE (2006) Discrete-event simulation of queues with spreadsheets: a teaching case. In: Perrone LF, Wieland FP, Liu J, Lawson BG, Nicol DM, Fujimoto RM (eds) Proceedings of the 2006 winter simulation conference

McKay MD, Conover WJ, Beckman RJ (1979) A comparison of three methods for selection values of input variables in the analysis of output from a computer code. Technometrics 22(v2): 239-245

Montgomery D (2008) Design and analysis of experiments, 7th edn. Wiley, New York

Miller JO, Bauer KW (1997) How common random numbers affect multinomial selection. In: Andradottir S, Healy KJ, Withers DH, Nelson BL (eds) Proceedings of the 1997 winter simulation conference

National Research Council (2008) Behavioral modeling and simulation: from Individuals to societies. National Academic Press, Washington

North MJ, Macal CM (2007) Managing business complexity: discovering strategic solutions with agent-based modeling and simulation. Oxford University Press, Oxford

Nelson BL (2008) The more plot: displaying measures of risk & error from simulation output. In: Mason SJ, Hill RR, Mönch L, Rose O, Jefferson T, Fowler JW (eds) Proceedings of the 2008 winter simulation conference

Nelson BL, Swann J, Goldsman D, Song W (2001) Simple procedures for selecting the best simulated system when the number of alternative is large. Oper Res 49:950–963

Netlogo website http://ccl.northwestern.edu/netlogo/

Ohno T, Bodek N (1988) Toyota production system: beyond large-scale production. Productivity Press, Tokyo

Opara-Nadi GE (2005) Electronic self-checkout system Vs Cashier operated system: a performance based comparative analysis. Capella University, Mississippi

Pidd M (2004) Computer simulation in management science, 5th edn. Wiley, New York

Press WH, Teukolsky SA, Vetterling WT, Flannery BP (2007) Numerical recipes 3rd edition: the art of scientific computing, 3rd edn. Cambridge University Press, Cambridge

Rinott Y (1978) On two-stage selection procedures and related probability inequalities. Commun Stat 7:799–811

Saliby E (1997) Descriptive sampling: an improvement over Latin hypercube sampling. In: Andradottir S, Healy KJ, Withers DH, Nelson BL (eds) Proceedings of the 1997 winter simulation conference

Sallach D, Macal C (2001) The simulation of social agents: an introduction. Special issue of social science computer review 19(v3):245–248

Samuelson DA (2000) Designing organizations. OR/MS Today December, USA

Samuelson DA (2005) Agents of Change. OR/MS Today February, USA

Samuelson DA, Macal CM (2006) Agent-based Modeling comes of age. OR/MS Today August, USA

Samuelson DA, Allen TT, Bernshteyn M (2007) The right not to wait. OR/MS Today December, USA

Sargent RG (2005) Validation and verification of simulation models. In: Ricki G. Ingalls, Manuel D. Rossetti, Jeffrey S. Smith, and Brett A. Peters (eds) Proceedings of the 2004 winter simulation conference 17-28

Schenk JR, Zheng N, Allen TT (2005) Multiple fidelity simulation optimization of hospital performance under high consequence event scenarios. In: Kuhl ME, Steiger NM, Armstrong FB, Joines JA (eds) Proceedings of the 2005 winter simulation conference

Srinivasan R (2002) Importance sampling—applications in communications and detection. Springer-Verlag, Berlin

Sullivan DW, Wilson JR (1989) Restricted subset selection procedures for simulation. Oper Res 37(v1):52–71

Sun R (2006) Cognition and multi-agent interaction: from cognitive modeling to social simulation, Cambridge University Press, New York

Swain JJ (2007) New Frontiers In Simulation: biennial survey of discrete-event simulation software tools. ORMS Today October, USA

Tang B (1993) OA-based Latin hypercubes. J Amer Statist Assoc 88:1392-1397

Williams EJ, Mohamed K, Lammers C (2002) Use of simulation to determine cashier staffing policy at a retail checkout. In: Verbraeck A, Krug W (eds) Proceedings 14th European simulation symposium

Wolff RW (1989) Stochastic modeling and the theory of Queues. International series in industrial and systems engineering. In: Fabrycky WJ, Mize JH (eds) Prentice-Hall, New Jersey

Zhao L, Sen S (2006) A comparison of sample-path based simulation-optimization and stochastic decomposition for multi-location trans-shipment problems. In: Perrone LF, Wieland FP, Liu J, Lawson BG, Nicol DM, Fujimoto RM, (eds) Proceedings of the 2006 winter simulation conference

Index